HISTORY OF THE CHILTERN LINE

JOHN M.C. HEALY

GREENWICH EDITIONS

CONTENTS

This edition published 1996 by
Greenwich Editions
Unit 7, 202-208 New North Road
London N1 7BJ

First published 1992

ISBN 0-86288-081-5

Printed and bound in Great Britain

CHILTERN LINE STATIONS

Name of Station	Notes	Date Opened	Date Closed
Marylebone		9 Mar 1899	—
To Aylesbury via Amersham			
Harrow on the Hill		2 Aug 1880	—
Moor Park	1	19 May 1910	—
Rickmansworth		1 Sep 1887	—
Chorleywood		8 Jul 1889	—
Chalfont & Latimer	2	8 Jul 1889	—
Amersham	3	1 Sep 1892	—
Great Missenden		1 Sep 1892	—
Wendover		1 Sep 1892	—
Stoke Mandeville		1 Sep 1892	—
Aylesbury	4DO	1 Oct 1863	—
To Banbury via High Wycombe			
Wembley Stadium	5	2 Apr 1906	—
Sudbury & Harrow Road		2 Apr 1906	—
Sudbury Hill, Harrow	6	2 Apr 1906	—
Northolt Park		1 Mar 1907	—
South Ruislip	7	1 May 1908	—
Ruislip Gardens	8	9 Jul 1934	21 Nov 1948
West Ruislip	9	2 Apr 1906	—
South Harefield Halt		24 Sep 1928	30 Sep 1931
Denham	10	2 Apr 1906	—
Denham Golf Links		22 Jul 1912	—
Gerrards Cross		2 Apr 1906	—
Seer Green	11	1 Jan 1915	—
Beaconsfield		2 Apr 1906	—
High Wycombe		1 Aug 1854	—
West Wycombe		1 Aug 1862	3 Nov 1958
Saunderton		1 Jul 1901	—
Princes Risborough		1 Aug 1862	—
Ilmer halt		1 Apr 1929	7 Jan 1963
Haddenham	12	2 Apr 1906	7 Jan 1963
Dorton Halt		26 Jun 1937	7 Jan 1963
Brill and Ludgershall		4 Apr 1910	7 Jan 1963
Blackthorn		4 Apr 1910	8 Jun 1953
Bicester North		4 Apr 1910	—
Ardley		4 Apr 1910	7 Jan 1963
Aynho park		1 Jul 1910	7 Jan 1963
Kings Sutton		1 Dec 1873	—
Banbury	G	2 Sep 1850	—
Grendon Underwood – Akeman Street			
Akeman Street	F	2 Apr 1906	7 Jul 1930

Princes Risborough to Aylesbury

Monks Risborough	13	11 Nov 1929	—
Little Kimble		1 Oct 1863	—
South Aylesbury Halt		13 Feb 1933	5 Jun 1967

Aylesbury – Bletchley

Quainton Road	P	23 Sep 1868	4 Mar 1963
Waddesdon Manor	14	23 Sep 1868	6 Jul 1936
Calvert	R	9 Mar 1899	4 Mar 1963
Claydon		20 May 1851	1 Jan 1968
Verney Junction		23 Sep 1868	1 Jan 1968
Winslow	E	20 May 1851	1 Jan 1968
Swanbourne		20 May 1851	1 Jan 1968
Bletchley (Oxford Plats)		9 Apr 1838	1 Jan 1968
Milton Keynes	E	20 May 1982	—

Neasden

Christian Salvesen	G	1 Apr 1990	—

Princes Risborough – Watlington

Bledlow Ridge Halt	–	15 Aug 1872	1 Jul 1957
Kingston Crossing Halt	–	15 Aug 1872	1 Jul 1957
Wainhill Halt	–	19 Jun 1925	1 Jul 1957
Chinnor	CP	15 Aug 1872	1 Jul 1957

Princes Risborough – Thame

Bledlow	-	1 Aug 1862	7 Jan 1963
Towersey Halt	-	5 Jun 1933	7 Jan 1963
Thame	O	1 Aug 1862	7 Jan 1963

Key

C = Cement trains (Chinnor branch now closed)
D = Depot
E = Used by excursion trains occasionally
F = Fertilizer
G = General goods
O= Oil trains (Thame branch now closed)
P = Preservation centre
R = Refuse trains

1) Formerly Sandy Lodge, now deleted from Network South East timetables.
2) Formerly Chalfont Road and junction for Chesham.
3) Originally Amersham and Chesham Bois.
4) Formerly Aylesbury Town.
5) Formerly Wembley Hill and Wembley Complex.
6) Formerly South Harrow and Roxeth.
7) Formerly Northolt Junction.
8) End of services marking handover to London Underground.
9) Formerly Ruislip and Ickenham.
10) Formerly Denham Junction for Uxbridge.
11) Formerly Beaconsfield Golf Links.
12) New station opened in 1987 as Haddenham and Thame Parkway.
13) Formerly known as Monks Risborough and Whiteleaf Halt.
14) Formerly Waddesdon

Stations listed above are those which were or still are on the present day Chiltern Line Network.

INTRODUCTION

The idea or mention of anything to do with the Chilterns usually conjures up in the mind a picture of rolling hills and lush green fields in an idyllic tranquil countryside surrounding close knit communities like Beaconsfield, Gerrards Cross, High Wycombe, West Wycombe, The Chalfonts, Great Missenden, Wendover, Little Kimble, Amersham, Chesham, Aylesbury, Banbury, Haddenham, Bicester, Thame, Princes Risborough, Chinnor and Watlington. Drawing on the rather beautiful, leisurely nature of life that is implied by references to the "Chilterns", the name has gained itself a new if unlikely importance, that of being a marketing propaganda term to attract potential customers to a little, under-used railway network that bursts forth from the western quarter of London and charges out through the dark satanic suburbs to Middlesex and into the very heart of Buckinghamshire and Oxfordshire.

Once upon a time this latter network, whose tracks only run as far as Aylesbury and Banbury today, belonged to a much larger system comprising more route miles. In its heyday this system played host to all different types of traffic, from quaint little trains trundling goods and passengers along branch or main lines acting as feeder services, to heavy expresses which made brief stops at important larger centres while on long distance journeys to the Midlands and the North.

Sadly, due to the escalating road construction programme which encouraged an increase in car ownership and the development of numerous road haulage firms, the majority of the aforementioned activity came to an all too abrupt end during the decade between 1960 and 1970, leaving stations and an infrastructure designed for greater things, to serve a quite basic commuter service. This was the way things stood until the early 1980's when once again the rapidly decaying lines from Marylebone down to Aylesbury and Banbury came under the intense scrutiny of the economists on the British Rail Board.

With the mention of reductions in services, rationalisation and rumours of closure of part of this archaic but quaint network being rife, a number of Rail Users and Campaigners were goaded into action to fight off the unthinkable, namely that these two humdrum commuter routes that not only had such an interesting background, but also had an important current role with great future potential, could pass away into history unnoticed. But at the eleventh hour despite the protests, two outside factors were to cause a reversal in the decision of the powers that be. The first was a huge increase in the popularity of the lines due to their promotion under the Chiltern banner and the appearance of steam trains which started running into and out of Marylebone again. Secondly the closure of Marylebone would have marred the impending launch of Network South East, the new operating region for railways in the London, Home Counties and South Coast areas.

In this book we shall see how the relevant railways in Middlesex, Buckinghamshire and Oxfordshire evolved, developed, operated and prospered until the 1960's when they were contracted to form the basis of what has now become a rejuvenated, modernised, upgraded smaller network with great potential, known as Chiltern Line.

John Healy (B.A.Hons.FRSA)

EARLY CHILTERN RAILWAYS

The two lines from Marylebone to Aylesbury and Banbury along with their associated freight branches which are operated by British Railways' Network South East and Railfreight Sectors under the Chiltern Line banner, would not have evolved had it not been for the coordinated plans of the myriad entrepreneurs who were active in creating companies to promote railways from the middle of the nineteenth century right up until the early part of the twentieth century. During this period known as the age of "Railway Mania", these interested parties who consisted of wealthy landowners and industrialists all over the country, put forward schemes to improve the conveyance of goods and to encourage the public to travel greater distances, but because of the grandiose nature of several of the plans, a drastically reduced number only ever came to fruition.

Thus the potentially profitable railway schemes began to take shape, gaining an immense amount of interest on the way, which helped their prosperity a result. The reputation and ability of these pioneer railways coupled to their possible expansion was recognised at an early stage. Consequently demand for railways continued to increase rapidly such that their construction was no longer confined to industrial areas only, but also extended to linking towns and cities of importance. Often many shorter lines were joined together to provide long distance routes, a fact which other promoters recognised as affording a wonderful opportunity for the building of branch feeder lines that would diverge from suitable existing or new interchange stations.

This was exactly how the Buckinghamshire part of the Chilterns came to have its first taste of the benefits of a railway, for as far back as the 1820's, George Stephenson had been surveying a possible course for his London and Birmingham Railway which would have followed a course similar to what became the bed of the Metropolitan/Great Central in years to come, running across from Uxbridge via the Wendover Gap, along to Aylesbury and then to Coventry before terminating in Birmingham. Naturally a plan to invade such idyllic countryside caused a vociferous protest from the local inhabitants and wealthy landowners who could see no immediate advantages existing for them. Stephenson bowed out gracefully and decided to plot a more easterly route which ran via Harrow, Watford, Berkhamsted, Tring, Cheddington, Rugby and Coventry. Consequently Aylesbury was precluded from being on the direct route, but it's new county town status, as well as being a terminal port on an arm of the Grand Union Canal, ensured it's importance was only ignored briefly, as a scheme was devised in the Autumn of 1835 to link it up with the London and Birmingham Railway. At this juncture Sir Harry Verney a prominent, wealthy landowner saw an excellent opening to help the Aylesbury Railway Company obtain an Act of Parliament for its railway in 1836 and to attract other new railways to the hitherto untapped Chilterns.

Once Parliament had given the Aylesbury Railway Company approval to construct the branch to Cheddington, numerous other entities investigated how they might perhaps take advantage of the scheme. One such was a consortium promoting a line over to Cheltenham via Aylesbury who successfully delayed the construction of the branch until the 12th May 1838. Nevertheless once work commenced on the seven miles of single track, supervised by Robert Stephenson, all efforts were made to complete the line quickly, which was helped by the light nature of the engineering required. Scarcely twelve months later the public opening took place on the 10th June 1839 amid loud cheers and shouts from the thousands of people who had assembled to watch the first trains arriving and departing as the Long Crendon Village Band provided a musical backdrop. Meanwhile a private banquet was held at the White Hart for the dignitaries. Such was the impetus of the occasion that 10th June was declared an official holiday for the future.

The line itself diverged from the London and Birmingham Railway at the Leighton Buzzard end of Cheddington station over a north facing junction, hugging the very sharply curved western edge of the main platform, before swinging round at an acute angle, after which it proceeded along a really gentle gradient on a virtually straight alignment to Aylesbury. About half way down was the only intermediate station on the branch, Marston Gate, situated in the middle of fields, which boasted an elaborate set of buildings but surprisingly basic goods facilities for the agricultural traffic it hoped to generate. Aylesbury however was totally different as it was provided with a superb track layout, in addition to a well appointed goods yard which echoed constantly to the rumblings of heavy agricultural traffic, especially milk. A small siding even led off into the nearby gas works which attracted a great amount of coal traffic. Passengerwise the station unfortunately was not too successful, probably due to the fact that the branch only had a north facing link to the main line at Cheddington and operated on a shuttle train basis with two locomotives plying their little carriages to and fro seventeen times a week in each direction.

The Aylesbury Railway initially displayed financial instability in the early years of operation, making it fall prey to takeover bids from other companies, but somehow the line remained as the the only link serving a large area of untapped territory between the GWR and L & BR. The latter fact did not elude the notice of a number of railway promoters who realised there was tremendous potential just waiting to be exploited. One such was the London, Worcester and South Staffordshire Railway who envisaged acquiring the branch to Aylesbury from where they would drive a line north westwards through Worcester as far as the Potteries. However the latter was an unsuccessful scheme, but it did not stop a second proposal from The Aylesbury and Thame Junction Railway Company, who tried fruitlessly, to encourage the extension of the branch along the edge of the Chilterns as far as Thame, but this also came to nothing. Failure of the A & TJR's plan undoubtedly came due to the London and Birmingham Railway conjuring up a counter scheme by backing the Cambridge and Oxford Railway Company who were desiring to connect the two University towns.

Part of this railway had it been successful, was to follow a course through Cheddington, Aylesbury and Thame. Now the Cambridge and Oxford railway believed they could achieve their aims, though they had underestimated their partners in the Varsity Line scheme, namely the L & BR who now announced a withdrawal of their support. They instead purchased the Aylesbury Branch on 16th July 1846, before laying down their own plans for an Oxford to Cambridge line.

Such activity in the middle part of Buckinghamshire had sparked off considerable railway interest in the south of the county due to the desire to widen employment opportunities which had been mainly restricted to small cottage industries or estate work and to curtail poor communications where journeys of thirty two miles could take twelve hours to complete. This meant that places like Ruislip, Denham, Gerrards Cross, Beaconsfield, High Wycombe, West Wycombe, Princes Risborough and Thame were all losing their trade importance and gradually fading into rurally insignificant areas.

Relief soon came to the ailing economy of the Chilterns with the proposal of a railway from Aylesbury to Harrow to Greenford and Wembley at which point it was to have joined up with the London and Birmingham Railway. However this line was in reality no more than a dream and it was left to the second proposal of the year to bring railways into the heart of the Chilterns. Under an Act of Parliament of the 27th July 1846, the Wycombe railway sought to establish a line from Maidenhead, on the Great Western London to Bristol Main Line, to High Wycombe.

With this scheme High Wycombe, which had been formerly described as an idyllic isolated location suitable only for people wanting to settle down in retirement to enjoy the rolling hills and deep dales through which a pleasant meandering brook flowed, was now destined to become a thriving centre especially with the rapidly expanding chair and furniture making industry. Thus a significant boost in trade and lifestyle of the local residents was expected but financial problems were to cause a halt in the construction of the railway. A strange anomaly had caused this setback which stated that the board of the Wycombe Railway Company were unable to sell off their land to the Great Western Railway for future development of the line, unless the company had spent more than half of its capital on construction works.

While legal wrangles were in progress over the creation of the link between Maidenhead and High Wycombe, a very sensible and potentially lucrative proposal was instigated which would have connected Southall to Hayes, Ruislip, Chalfont, Beaconsfield, High Wycombe, Aylesbury and Cheddington. This scheme, promoted by Isambard Kingdom Brunel would have not only provided a link between the GWR and The L & BR serving many important places en route but also allowed an interesting variety of inter-regional long and short distance through train travelling.

Unbelievably though by a strange quirk of fate which altered the course of railway history drastically, the 1847 Great Western to London and Birmingham Railway link was abandoned and the Wycombe Railway survived to fight another day. Not to be outdone Isambard Kingdom Brunel desirous of a challenge, turned his attentions to rescuing the ailing Wycombe Railway project, with the belief that he could effect a GWR to L & BR link by eventually extending the Wycombe Railway as far as Aylesbury where it could join up with the branch to Cheddington. However the first task was to resolve the financial problems which the Wycombe Railway Company had been experiencing, a task soon achieved with such a notable engineer at the helm. Thus construction was soon able to be resumed in the early 1850's, some six years after the act authorising the line was first passed.

Sadly fate was to intervene again when the line's contractor was declared bankrupt and a further lack of finance was experienced. As soon as this news was broadcast, it was not long before a new company tried to make capital out of the unfortunate situation. This was the London and Mid Western Railway who devised a route that would push the Oxford, Worcester and Wolverhampton Railway south eastwards to Princes Risborough, High Wycombe and Uxbridge before heading to Brentford from where it would link up with the London North Western Railway. Despite the grandiose plans of the latter, the scheme was a nine minute wonder and soon the Wycombe Railway was back on its feet such that construction could begin once again in the Summer of 1853.

In a little over a year on August 1st 1854, much to the delight of the local inhabitants, the first goods and passenger traffic was carried on the single broad gauge track laid with 90.5lb and 12in width base Barlow Rail which extended from Wycombe Junction just to the west of Maidenhead Station and ran alongside the Wye Valley terminating close to the heart of High Wycombe. Numerous wayside stations with lavish facilities were erected at Cookham, Marlow Road (later renamed Bourne End), Wooburn Green and lastly Loudwater (where a goods passing loop was laid). In later years to cater for the rapid sprawl of new housing around Maidenhead, a small halt was also established at Furze Platt, opening on the 5th July 1937.

Although the Wycombe Railway was a resounding success, a period of almost three years elapsed before any further enlargement of the railway network in the Chilterns took place. Finally on the 17th August 1857, Parliament passed an act paving the way for an extension of the Wycombe Railway out as far as Thame. Indication of a firm commitment of this kind pleased the inhabitants around Princes Risborough and Thame greatly, despite the fact that their aim of boosting the local economy via a new railway had so often been frustrated with promoters offering various grandiose schemes which promised considerable benefits which not surprisingly came to nought. Additionally the Wycombe Railway proudly offered some extra substance to it's investors as the company promised to pay out a large four per cent dividend to its shareholders once the new line opened for business.

As far as capital outlay was concerned for the new High Wycombe to Thame line, an estimated £400,000 was deemed to be sufficient to cover the complete cost of all the engineering works such as cuttings, embankments, stations, track, ballasting and signals over the thirty six miles. Half the aforementioned sum would be provided by the Great Western Railway who would run the line on it's completion, while the remainder would be raised by selling 40,000 shares worth £10 which each bore a £1 deposit. With such a task ahead of them, the Wycombe Railway's principal officials rapidly rounded up various wealthy, influential individuals to attend a meeting held at the Spread Eagle Hotel in Thame. A very mixed reaction greeted the proposed scheme with some investors readily offering capital while others were more cautious, with the result that insufficient finance was raised. Following the meeting some convincing arguments won over the waverers, but even then they were still short of the total, when suddenly a clever idea was conjured up by the company's solicitors Messrs Baxter Rose and Norton, who resolved that Thomas Tredwell and Company the chosen contractor, should be entitled to possess a £30,000 holding as part payment for all the construction works required.

Thomas Tredwell was responsible for providing all the necessary materials and labour, but the Wycombe Railway Company dealt with acquiring the land which was in the shape of a long narrow strip thirty six miles long, running through open countryside on a not too harsh terrain which would make for light engineering works. Engineeringwise the Wycombe Railway would have been fortunate in having the services of the great Isambard Kingdom Brunel who was to supervise construction of the new line, had he not died. Thus the post of Chief Engineer fell vacant. However help was soon to come as his very capable understudy E.F. Murray draughted a rather obsequious letter to the Wycombe Railway Company board which read as follows:

"I would beg leave to state that for quite some time prior to Mr. Brunel's death, I was aware that it was probable he might decline to continue to act as engineer of the line, and I was also aware of his very kind intention to recommend me to the board for it's favourable consideration as his successor. If he had lived, it is more than possible that such an arrangement would have been proposed, and that I should have had the great advantage of his personal advocacy and support."

"But however valuable that has been and however gratifying the knowledge of Mr. Brunel's favourable opinion, after experience of 22 years I would venture with great respect to solicit your consideration of my claim to the high honour of being appointed your engineer on other grounds also. For many years I have had the honour of acting as your resident engineer, not only on the original line but also on the extension … and I have always earnestly striven to forward the interests as well as promoting the prosperity of the company … If you consider me to be worthy of your confidence it will be my constant study to endeavour to deserve it".

Although rather a long letter, Murray had a definite way with words such that he came across as being a modest, capable man even if he displayed the signs of being an "Ambitious Creeper" in modern parlance. Such self advertisement proved to be the right formula, as the Directors prompted by the letter, after due consideration at an emergency board meeting of the Wycombe Railway, installed Murray as the new engineer.

While Murray had been securing his new post, meanwhile at a small ceremony in September 1859, accompanied by a customary dinner and speeches, Edward Griffin one of the Directors of the Railway who owned Towersey Manor, ceremoniously cut the first turf since the line was passing through his land. Soon after this really modest start had been made, work began to progress so rapidly, that by the 2nd February 1861, Murray informed the Directors that apart from one earthwork, the whole series of embankments and cuttings had been finished from High Wycombe to West Wycombe, Saunderton and Princes Risborough up to Bledlow. In addition from Towersey to Thame several girder and brick bridges were ready, along with crossings, some of the embankments and cuttings and a number of culverts. Consequently this just left construction of the final length from Bledlow to Towersey which encountered a brief hold up through difficulties in acquiring the land, but fortunately this stretch required only light engineering.

Once the trackbed was complete, the laying of the permanent way and ballasting could begin. To obtain ballast easily, Murray had purchased a field from a Mr Rogers adjoining the Wycombe Railway at Cookham, so that his contractors could extract the materials as required and transport them to the site by train. However the latter process was considerably hindered by the onset of terribly inclement weather, a fact which made Murray edgy to the point of providing regular reports to the Directors, even when he did not have anything to tell them.

The inclement weather did have one benefit though in as much as it proved to the Directors that the bridges, viaducts, cuttings and embankments could stand up to the elements. While a content Murray battled with the weather to complete the Thame extension, the Directors of the Wycombe Railway became eager to extend the line to Oxford and establish a branch from Princes Risborough to Aylesbury. Their move was largely prompted by the fact that they wanted to take advantage of the support they had from the Great Western Railway who not only injected substantial capital into the Maidenhead to High Wycombe line but also promised a nucleus of rolling stock to operate it, with a similar provision for the Thame extension.

Feelings by this time were again running high that they should be making further inroads into surrounding railway free zones, before any rival schemes appeared and came to fruition. Indeed the West Midland Railway Company (who owned the Oxford Worcester and Wolverhampton Railway) attempted to do just that by introducing a parallel proposal to establish a railway from a place to the north of Oxford known as Yarnton to Thame thence to Princes Risborough before heading via Amersham, Beaconsfield and Uxbridge into a London terminus at Knightsbridge. Also to be incorporated was a branch to Aylesbury and a connection to the West London Line at Shepherds Bush.

So similar in nature was the scheme that it caused an enormous outrage, encouraged by a sea of correspondence flowing between the West Midland Railway's William Fenton and Lord Shelburne of the Great Western Railway. A fair solution could not be arrived at to resolve the situation as to whose company was ultimately going to benefit. Consequently a court case ensued between the two parties, but events took a surprising turn when on the day legal proceedings were about to commence, an announcement was made to say that the West Midland Railway would be amalgamated with the Great Western, though the reasons were unclear.

Thus the way ahead lay clear for the Wycombe Railway to proceed unhindered allowing a smooth passage through Parliament on 28th June 1861, of the Bill for the Aylesbury and Oxford extensions. Murray was delighted at the faith the directors had shown in him such that he speeded up the schedule for completing the line to Thame, as well as engaging in numerous inspections and train trials before the opening on 1st August 1862. This was performed without much pomp and circumstance with an engine hauling three carriages into Thame station at 10a.m. carrying several station officials. A little later in the day a large party of directors, shareholders and members of the public gathered for the opening ceremony. After this at about 3.30p.m. the first train left for Paddington. On the next day a regular service of four trains a day in each direction began.

Soon afterwards Murray set to work on his next task, namely the short branch from Princes Risborough to Aylesbury which required scarcely any mammoth engineering works as it ran mainly along a plain through Monks Risborough and Little Kimble. Finally on 1st October 1863 Aylesbury gained its second railway some twenty four years after the London and Birmingham Railway had penetrated the town. Unlike the branch to Thame, the spur to Aylesbury operated on a shuttle basis from Princes Risborough.

Subsequent Wycombe Railway Co Developments

1864	Bill For Railway from Loudwater to Uxbridge.
24 Oct 1864	Opening of the Thame to Oxford Railway.
1 Feb 1867	Absorbing of the Wycombe Railway into the Great Western Railway company.
23 Aug 1870	Authorising and instigating of a gauge change from the Broad to Narrow Gauge of 4ft 8,1/2ins.
1 Sep 1870	Completion of above works.
1 Oct 1872	Woburn Green renamed Wooburn Green.
28 Jun 1873	Great Marlow Railway Opened from Bourne End to Marlow.
1 Jan 1874	Marlow Road renamed Bourne End

All this surge of activity during the afore-mentioned decade had not only brought success for the companies involved, but it also whetted the appetite of several other bodies equally desirous of bringing railways to as yet unconquered Chiltern territory. Alas ardent enthusiasm got the better of practicality such that in the end only some were successful. Those which failed were as follows and in the main did so through being too ambitious.

1864	The London and Rickmansworth Railway, which would have started from the LNWR at Stonebridge Park and run via Alperton (with a branch to the G.W.R main line through Greenford), Roxeth, Pinner etc. This would have linked up with a proposed Amersham and Chesham Railway.
1865	The Harrow, London, Buckinghamshire and East Gloucester Railway. There were to be connections at Cheltenham for South Wales to lure coal traffic away from the G.W.R.
1871	A line from the LNWR at Wembley running via Greenford, Ruislip (with a branch to Uxbridge), Gerrards Cross, Beaconsfield, Wycombe and across the Vale of Aylesbury.
1873	A revision of the former plan that altered its course to follow that later used by the 1899 G.W./G.C. route.

From the above proposals it was patently obvious that there was still a strong bias towards laying a railway across the Vale of Aylesbury, an aim which had been in the minds of many since the 1840's. Undoubtedly the most prominent figure with this aim was Sir Harry Verney who had been closely involved with the arrival of the Aylesbury Railway in 1839 and his close friend the Duke of Buckingham. They had each proposed schemes for an Oxford and Bletchley Junction Railway and Buckingham and Brackley Junction Railways respectively, undoubtedly because they wished to serve their estates. Both projects had considerable support from the London and Birmingham Railway whose line they would effect an interchange with at Bletchley. However some major changes were to take place before construction began, namely that the London and Birmingham Railway and Grand Junction Railways amalgamated to become the London North Western Railway in 1846, insisting that the two undertakings of Sir Harry Verney and the Duke of Buckingham (who handed over the chairmanship to his friend due to severe financial pressures) merged to become one under the appropriate banner of the Buckinghamshire Railway. Consequently there was much rationalising of the initially envisaged amount of permanent way which did away with the proposed southern arms to Aylesbury, Tring and Harrow. Once this had been effected, a bill was put forward before Parliament to approve the scheme.

By mid 1847 the Buckinghamshire Railway received a green light for its new lines and consequently appointed Thomas Brassey, an engineer with no less an ability than Robert Stephenson. He soon set about mustering a huge work force from the local inhabitants thus bringing much activity to an otherwise sleepy area. Unlike a number of other concerns, the Buckinghamshire Railway Company demanded that contractors provide not only proper accommodation and regular pay for the navvies but also to ensure their unruly behaviour as a result of drinking too much was checked by police vigilance and the auspices of a chaplain. Soon these measures had been catered for, such that at 5pm on Tuesday 20th April 1847 by the Cross Keys public house in the Landborough Road Buckingham, construction began with the cutting of the first sod. Initially work started on both the Oxford and Banbury lines, and advanced steadily until suddenly in 1849 it came to the attention of Sir Edward Watkin, the Buckinghamshire Railway secretary, that some drastic economies needed to be effected due to an impending lack of finance. His quick realisation of the situation saved the day and to combat the problem, he recommended a temporary suspension of work on the Oxford line, reducing the Banbury Branch to single track saving nearly £90,000 and cutting out two proposed stations at Padbury and Finmere, leaving Swanbourne, Winslow, Buckingham, Brackley and Farthinghoe as revenue earning points which cost a total of £23,529 to build. As for the actual line from Bletchley to Banbury this required a sum of £440,600 to effect a sufficient state of completion ready for the events detailed below:

Date of Event	**Description**
26 Mar 1850	Inspection of Banbury Branch by Directors, Engineers and Traffic Managers on board a special train followed by a lavish dinner at Winslow in the afternoon.
19 Apr 1850	Railway Inspectorate Examination of the Branch by a special train consisting of two engines and tenders with Captain W. Wynne on board. With few exceptions he deemed the work to be satisfactory.
23 Apr 1850	Permission withheld for opening until the completion of the sidings was effected, with particular reference to points and crossings at Banbury Station.
29 Apr 1850	Examination by Railway Inspector Captain Laffan of the points and sidings.

1 May 1850	Opening of the Buckinghamshire Railway in the early morning with a large multitude gathering to welcome the new line as well as watch the four departures of the day at 6.30am, 9.45am, 1.45pm and 5.00pm, which formed the regular Buckingham to Bletchley service operating times.
15 May 1850	The opening of the Branch to goods traffic even though the facilities were not really ready.

In the ensuing months once operations were running smoothly on the Banbury Branch, all effort could be directed to opening the Oxford line, though it had to terminate at Islip almost sixteen miles west of Winslow where it diverged from the former. Beyond Islip the railway could only be finished with difficulty because much indecision along with the bandying about of some ridiculous schemes regarding the line's entry into Oxford so it offered the maximum operating potential, hampered progress. Eventually three varying courses were outlined which are shown below:

Route 1	From Islip the line would follow the River Cherwell then cut through the village of Summertown, from where it was planned to cross the Woodstock Road and Oxford Canal up to a complex junction with the new Oxford, Worcester and Wolverhampton Railway. This would require a mixed gauge track from the latter point as far as Oxford.
Route 2	Linked into one of the three abortive London Oxford and Cheltenham Railway, Tring to Oxford Railway and London, Bucks and East Gloucestershire Railway proposals, it would have skirted around Summertown, prior to following the Banbury Road and sweeping round a three furlong curve, after which it was to cross the river and the Banbury and Woodstock Roads. Then it would head for a junction with any one of the three aforementioned schemes near Wolvercote.
Route 3	From Islip this route went to the west of Summertown, then penetrated Water Eaton, Wolvercote and Port Meadow as far as the edge of Port Meadow to Botley Road.

The first course failed because of the Buckinghamshire Railway's connections with the London North Western Railway in whose fold was the former London and Birmingham Railway company. The latter organisation became embroiled in an acrimonious dispute with the Great Western Railway in 1845, concerning which company would be able to extend north from Oxford to Birmingham. After a month's long hard struggle involving many famous names and dignitaries, the Great Western won the day. However the London North Western Railway persisted in their quest, thinking they could penetrate Oxford by invoking the "Mixed Gauge" agreement which meant they could lay a third rail within the broad gauge. The prospect of any company linked up to the old London and Birmingham Railway, using GWR metals or stations was unthinkable.

In the case of the second route, failure came from it being based around three main line route proposals with little substance that involved some rather bizarre engineering works. Unlike the first scheme, several facts became apparent in that the Buckinghamshire Railway would have to run independently into Oxford, so dictating the need for an expensive terminus. From the latter a third plan emerged which was theoretically sound, but practically difficult to effect as it had to be carried out in three stages. Part one, sanctioned by a Parliamentary act in 1849, took it to a temporary terminus at Oxford Road a year later. A permanent site closer to the centre of Oxford did not become available until Christ Church College advertised their desire to dispose of land then occupied by the ruined Rewley Abbey. In theory this was all very well, but to join up the temporary terminus and the proposed permanent one, the railway would have to cut across Port Meadow which came under the jurisdiction of the Freedmen of Oxford. Consequently a series of delicate negotiations ensued with the Freedmen, who in the end resolved to accept compensation from the railway. Having overcome this hurdle, work progressed rapidly such that by early 1851, the way was clear for running test trains. By the 20th May the route was deemed satisfactory and duly opened from Bletchley to Oxford. Coincidentally a banquet was held at which some directors of the London North Western Railway agreed to provide motive power and rolling stock for the Buckinghamshire Railway's train services. Indeed the new line enjoyed such success that within just three years, a second track was laid between Oxford and Claydon, along with the upgrading of all the stations and their goods facilities costing some £20,500.

Thus the Buckinghamshire Railway managed to achieve part of it's aims for the moment, but within the Parliamentary Act authorising the latter network, as stated previously, there were clauses for extending from Claydon to Aylesbury, Aylesbury across to Tring and Aylesbury up to Harrow. The latter two were actually dropped due to a mixture of the Harrow scheme being too costly whilst the one to Tring would have duplicated the already existing Aylesbury to Cheddington line. As for constructing a new stretch of railway from Claydon to Aylesbury, this annoyed the London North Western Railway who leased and also operated the smaller Buckinghamshire concern to the extent that they opposed the idea vigorously.

However such action was regarded as sheer folly by several local parties and trades people who were bursting to see Central North Bucks opened up by the arrival of another new railway, but it was not until some positive action was taken by a Chronicler Mr J.K. Fowler who, having met up with a Railway Engineer Mr. Brydone and a Builder Mr.F. Rummens, took the assembled party by appointment to see Sir Harry Verney. Sir Harry was absolutely delighted with the interest they had shown and after pledging his full support, urged the three to make a rapid deputation to his old friend the Duke of Buckingham. Fortunately the Duke was in residence and he was equally enthusiastic to the point of offering financial help of £5000 along with an assurance that his position as the London North Western Railway Chairman would not only ensure approval of his company but also its agreement to operate the new Aylesbury and Buckingham Railway once it opened. Needless to say there were various strings attached, namely that he became the Chairman and Sir Harry Verney was made Deputy Chairman, that the route was to run along a more circuitous course to the west of Pitchcott Gap so he could build a railway into his estate at a later date and that it served Sir Harry Verney's estate at Claydon.

Slightly to the East of Claydon was the point where the proposed new line was destined to part company with the Oxford and Banbury portions of the Buckinghamshire Railway. Here an interchange was to be set up appropriately called Verney Junction. Leaving Verney Junction the line would proceed more or less southwards up to the Winslow Road where the first intermediate station would be sited. From Winslow Road the contractors encountered a very sharp curve up to the crossing of the Grandborough Road, near to the location of the second station. Soon after the route followed a more south westerly direction so avoiding Quainton Hill. Quainton itself was reached after negotiating a further sharp curve and here a third station was planned which was ideally placed to form the junction with the Duke of Buckingham's proposed Wotton Estate Tramway. An almost straight alignment through verdant countryside brought the railway then into the north west quarter of Aylesbury. From here under a second proposal, the Aylesbury and Buckingham Railway was to head a little to the south of a revamped Town station, forging a link with the Aylesbury to Cheddington Branch over a new chord. Sadly the construction of the new line was to be far from smooth and its completion proved very difficult as can be observed from the sequence of events below.

Jan	1860	Initial Discussions. F.Rummens appointed as builder.
Aug	1860	Royal Assent for the line given.
Feb	1861	Construction began on the line from the Claydon end, initiated on a capital of £98,000 in 3290 £25 shares. Departure of the Duke of Buckingham from the board of the London North Western Railway.
Jul	1864	Capital increased by £12,000 in an act of Parliament.
Jun	1865	Capital increased by a further £111,000 by sanction of Parliament with the stipulation that the line must be completed within five years
???	1866	Investment crisis and problems caused by landowners seeking compensation from the new railway.
???	1867	Work began on the Aylesbury and Buckingham Railway to the London North Western Railway link but sadly the London North Western Railway soon changed their policy making a working agreement with the Aylesbury and Buckingham Railway impossible.

The aforementioned deed was a dastardly stab in the back which had largely occurred not only because the Duke of Buckingham no longer held a powerful position on the board of the London North Western Railway but also because the company secretary had died. With this the Aylesbury and Buckingham Railway would have been left out in the cold without any stock to operate trains on its new line, a situation which both the Duke of Buckingham and Sir Harry Verney would not allow to prevail, so they turned to the Great Western Railway for support using their remarkable skill in railway negotiations. Surprisingly the Great Western Railway were all too eager to enter in to an operating agreement with the Aylesbury and Buckingham Railway, because not only would the former be able to extend their territory but it also meant they could get back at the London North Western Railway for a number of previous episodes where battle lines had been drawn up.

Amazingly after the passage of eight years, the line and the new stations at Aylesbury, Quainton, Winslow Road, Granborough Road and Verney Junction were finally completed ready for their grand public opening on 23rd September 1868. Services were provided by ex Oxford Worcester and Wolverhampton Railway stock absorbed into the Great Western Railway fleet comprising of two locomotives and three coaches which trundled gently along the single line in each direction calling at all stations three times daily. This pattern remained with a prolific amount of goods being handled, but sadly the line's complete potential was not realised since the ultimate destination of Buckingham was never reached probably due to lack of finance brought about by the fact the Aylesbury and Buckingham Railway were required to supply some £28,000 to allow the Great Western Railway to convert their Aylesbury Branch from Broad to Narrow Gauge, which would enable through working from Paddington to the wilds of Buckinghamshire, however this facility was never taken advantage of.

Almost a year after the opening of the Aylesbury and Buckingham Railway, on the 27th July 1869 with backing mainly from the Earl of Macclesfield, an act was passed authorizing the construction of a railway between Watlington and Princes Risborough. This new line ran out of Princes Risborough on a parallel course with the Thame branch before diverging in a south westerly station to the outskirts of Bledlow where a simple halt was erected. On leaving Bledlow while skirting the foot of the Chiltern Hills, the route made a gradual ascent of 1 in 68 until it crossed Wainhill Road where a Halt was later established in 1925. A short distance to the south lay one of the two principal intermediate stations at Chinnor which boasted a single platform with some ornate red brick buildings on it and a small goods yard accommodating just two sidings. These were augmented in 1927 by a network of track to serve a massive lime and cement works further down the line. Falling away from Chinnor the new railway headed on a downward gradient past the future site of Kingston Crossing Halt up to the next significant port of call at Aston Rowant, which contained a similar set of facilities to those at Chinnor. Beyond Aston Rowant the engineering works were extremely light indeed right through Lewknor where a halt was built, to the lavish terminus at Watlington. Here the contractors created a short platform, a corrugated iron shed capable of holding an engine and a trailer car, a goods yard and three sidings. Due to the light nature of the total works the line only took a period of over two years to complete and opened on 15th August 1872 with a service of some five passenger trains each way, every day with an extra working on Saturday evening and Monday mornings. These were supplemented by three Goods trains over the branch.

A few months prior to the Earl of Macclesfield's railway being completed, another Local Landowner saw his dream of building a line through his own estate come to fruition. This was the Duke of Buckingham who was encouraged by the fact that a railway now passed through Quainton which was not all that far from his seat at Wotton. The Duke was very fortunate in that the course of the tramway would mainly run over his own land, so permission did not need to be obtained via the rather complex Parliamentary process as was usual. However agreement needed to be reached over the use of two short stretches of land, belonging to the Winwood Charity and Baron Ferdinand de Rothschild who lived in Waddesdon Manor, before the railway could be built.

Even though the new line was designated a Tramway, it was to be constructed to standard gauge, leaving Quainton Road in a south westerly direction whence it ran parallel to the road up as far as its crossing of what is now the A41. After the A41 crossing the course pursued a gentle curve as it skirted the western side of Westcott village from which point a branch was later included to serve Waddesdon Manor Gas Works. A little further on the line crossed the Westcott to

Ashendon Road before running in an almost straight line up to Wotton bisecting one of the driveways leading up to Wotton house. From here the tramway curved into the village of Wotton Underwood where the proposed terminus was to be sited. In order to collect goods and minerals from the various important points on the estate, two primitive platforms made of heaped up earth retained by sods would be established en route at Waddesdon and Westcott along with sidings for wagons.

Soon afterwards construction of the line began on 8th September 1870, making full use of the estate workers whose progress was really rapid, since the line on the whole ran on level, fertile arable land that required little or no major earthworks, under or over bridges, culverts, or heavy drainage channels. Once the base was prepared, the line was ready for ballasting but this presented problems as the Duke could only obtain materials for the 6" of top ballast formed from burnt clay. The 9" of base materials consisting of hard gravel had to be shipped at great cost to the Tramway from Buckingham via the London North Western and the Aylesbury and Buckingham Railway.

When the ballast had been laid and settled, a firm of contractors Messrs Lawford and Haughton of Westminster Chambers, Victoria St, London were draughted in for the specialist task of securing the sleepers and rails in position. The sleepers from large Norwegian trees were sawn into 6"x6" widths, before being creosoted with approximately 8–10lbs of oil per cubic foot and finally they were positioned lengthways on the trackbed and linked together at the ends by means of two flat iron staples being driven in. To ensure that the gauge was maintained throughout the contractors inserted numerous 4'8" width cross struts between the longitudinal members fabricated out of waste timber scavenged from the Duke's estate. Extra measures were taken to ensure the latter structure remained in place by attaching the two sets of longitudinals together with iron bars. Having done all this the contractors were able to lay the 21ft long 30lb bridge rails, though this was not without one or two problems, since the sleepers had been too heavily treated to toughen the wood such that it made insert of the rail-holding fang bolts exceedingly hard. At each end the latter were linked with an eight inch fishplate tucked into the recess of the rails. With this kind of construction the track would only bear limited weights and allow trains to operate at very low speeds.

With both the efforts of the estate workers and the contractors, the Quainton to Wotton tramway project came to fruition speedily such that it was able to be opened to goods and mineral traffic with a small ceremony on 1st April 1871, just seven months from the time construction had begun. So absolutely ecstatic was the Duke of Buckingham that he immediately commissioned an extension of the tramway which would diverge at Church Siding, heading due south westwards until it bisected the driveway from Thame Lodge to Wotton House before assuming a parallel course to the Brill Road up to the point where it encountered the Ludgershall Road. Finally it would drop down into a little field which lay at the bottom of the hill some 3/4 mile away from Brill village. While the people of Brill were delighted that the Duke had thought fit to bring a railway to their community, they felt the picture was not quite complete without a passenger service. To illustrate the point some one hundred and fifty excursionists decided to travel on a special train to London from the branch on 21st August 1871, the same day as the commissioning of Wood Siding station, on the new section.

At around this time good progress was being made on the remainder of the extension towards Brill despite numerous severe gradients over the 2.75 mile distance. These ranged from 1:100 to 1:51 with the line rising some 130 feet altogether from Wotton. Within three months the next notable point in the line's construction was reached on the 29th November namely Cottons Lane Gate. A siding was laid here primarily for the benefit of the local brickworks. Returning to the main line the last quarter of a mile into the station at Brill did not come into use until March 1872.

Naturally the Duke was jubilant at the opening of the full length of the Tramway, but soon he became eager to increase the earning potential and usefulness of the line by employing his work force on the construction of a further offshoot. The latter would leave via a junction close to Wotton Station, then proceed to Kingswood Lane, where the estate had a brick and tile works. Eventually the lightly constructed tramway would terminate some 1 mile 57 chains from its origin with a coal wharf served by two sidings, a small turntable and a staff dwelling.

METROLAND AND THE LAST MAIN LINES

Following the completion of the Wotton Tramway or Brill Branch, the development of the railways in the Chilterns was to take an interesting turn, mainly due to the appearance of an entrepreneur from the north called Sir Edward Watkin, who was the enterprising son of a wealthy cotton merchant. Already he had made a name for himself since he had been employed as Chairman of the Manchester Sheffield and Lincolnshire Railway for nearly eight years. His reputation then won him the opportunity to be offered the Chairmanship of what was a very disorganised and financially unstable Metropolitan Railway. Naturally Sir Edward Watkin, with his thirst for a challenge, jumped at this chance, especially as this position could later enable him to fulfil a substantial part of his aim.

The latter aim which had it come to fruition would have changed the course of railway history, was to create a line running from Manchester, Sheffield and Lincoln and other industrial cities of the north to London, whence it would connect with the East London Railway, before heading onto the South Eastern Railway to Dover, at which point it would disappear into a tunnel under the Channel until it emerged on the continent. Such a concept had magnificent potential and Watkin realising this resolved to act with extreme caution, by edging southwards with the Manchester, Sheffield and Lincolnshire Railway to a point where it could effect a junction with the Great Northern near Annesley to the north of Nottingham. Watkin firmly believed that he could draw up a working agreement to use either the latter's line or the neighbouring Midland Main Line to give his trains access into London from where he already had a direct route to the Channel via the East London and South Eastern Railway's which he held the chairmanship's of. Indeed it was no surprise that neither the Great Northern Railway nor the Midland Railway could reach an agreement with Watkin since they were not really eager to have another company's traffic running on their lines especially if there was an element of competition. So Watkin had lost his way for the moment and whilst he prepared a new strategy, he kept the Metropolitan shareholders in the dark about his larger project for the time being.

Soon once the inner city lines had managed to show their success, the Metropolitan began to investigate the possibilities of a long extension into the suburbs to attract other potential custom. In hardly any time at all, the necessary Parliamentary Approval came allowing construction works to be undertaken from the then limit of operation at St.John's Wood and Swiss Cottage. From Finchley Road the railway ran above ground as far as Willesden Green which was reached in 1879. Willesden essentially became an intermediate stage, since the primary objective was Harrow, where a fine Queen Anne Style Station was erected with sumptuous facilities and even stained glass windows. Opened on the 2nd August 1880, the station offered some thirty six trains daily with fewer on Sundays. Thus Harrow now was becoming a popular commuter belt and indeed demand arose for the railway to head further Northwards.

In 1874 an earlier proposal to build the Harrow and Rickmansworth Railway was incorporated by the Metropolitan Railway for a future development. However the opportunity was not exercised until 1881 when contractors started to push the line along via North Harrow to Pinner initially which was completed in 1885. At this juncture progress was being made with the Rickmansworth section and Sandy Lodge station such that almost two years later this was added to an already rapidly expanding line.

The Metropolitan Railway company were naturally pleased with the results, especially as many of the local residents were attracted to make heavy use of their new transport facility. Fired by such success, Sir Edward Watkin convinced his shareholders that it was in their interests if the Metropolitan set its sights on Chesham which was a place of significance offering benefits in terms of freight traffic from the local breweries, woodware, boot making, shoe making, brush, straw plaiting and flour milling industries. The inhabitants of the Chesham who believed they would not enjoy the privileges of their own station, now became very interested in the railway and pressed for construction to commence on their line almost immediately after the opening of Rickmansworth on the 1st September 1887. In order to give the residents an opportunity to make use of the Metropolitan, Watkin established a temporary horse bus service between Chesham and Rickmansworth, which began its journey from the site of the new terminus which lay in Mill Fields on the outskirts of town. The latter prospect considerably annoyed the locals who raised some £2000 to purchase land leading to a location a half a mile further north. The Metropolitan made its response by buying a further piece of land to establish what was a large commodious goods yard. Eventually the happy day came on the 8th July 1889 when Chesham station opened to the public, offering a journey time of 50 minutes to Baker Street. Needless to say the Metropolitan were not content with this achievement and within a few months sought to extend the line to Tring where it was hoped they could form a link with the London North Western Railway. Despite a portion of one and a half miles of land being purchased, the scheme fizzled out.

Although the failure of the Tring extension was obviously really demoralising, the whole affair seemed to be only a minor setback compared with the problems emerging for Sir Edward Watkin on his Channel Tunnel scheme. Financing of the tunnel itself was a major source of concern and now in order to complete the scheme, Watkin would have be forced to construct a new main line from Sheffield southwards. At this juncture he realised that if it was to follow a course through the centre of England, through Nottinghamshire, Leicestershire, Warwickshire, Northamptonshire and head down as far as North Buckinghamshire, there would be a great opportunity to effect access to London Baker Street. However a direct run of this nature would only be possible if the separate stretches from Quainton Road to Aylesbury Town and Chalfont Road to Baker Street were linked with a new section of line.

Having reflected on all the implications and the benefits of his plan, Sir Edward Watkin lost hardly any time at all in arranging a meeting in 1890 with the shareholders of the Metropolitan. At this gathering he made it abundantly clear that they were going to be offered a "Valuable Present" which would bring their lines considerable extra traffic and a welcome prosperity on condition that they join forces with the larger Manchester, Sheffield and Lincolnshire Railway. Mind you the Metropolitan were not wholly convinced that merging would be in their interests especially as there was doubt over the soundness of the Manchester, Sheffield and Lincolnshire Railway's finances which earned them the title the "Money Sunk and Lost Railway". Eventually though the latter were offered running powers instead, leaving the Metropolitan to take advantage of the situation by first of all taking over the little Aylesbury and Buckingham Railway on the 1st July 1891 and secondly by their extremely rapid completion of the new Chalfont Road to Aylesbury line, which had a temporary terminus at Brook Street when it opened on the 1st September 1892. Afterwards the Great Western Railway installed a new junction off their branch to Princes Risborough which allowed Metropolitan trains to work into the joint station which had been suitably modified for the extra traffic.

While all this activity was taking place in the South, progress was being made in the north to bring the Manchester, Sheffield and Lincolnshire Railway from Beighton in South Yorkshire down to Annesley in Nottinghamshire from where the London Extension would spring. During this time a Bill was put before Parliament for the Annesley to Quainton Road section. Not surprisingly in its first reading several objections were raised from Watkin's rivals and other interested parties like the Marylebone Cricket Club, the London Council and the large colony of artists around Swiss Cottage and St.John's Wood. Consequently the proposal was rejected by Parliament, but Watkin refused to be dogged by one more failure, so he resolved to try again after making several necessary amendments to his scheme and to provide evidence of the considerable support that he had from traders and a number of wealthy backers.

Finally on the 28th March 1893, he was successful. However the delay increased the cost of constructing this already expensive line still further. When the contracts were placed, costs were running at £3,182,155 without any signalling, rolling stock or rails included. For this an extra £6,200,000 of stock actually had to be issued at various stages throughout the construction, along with some £2,066,666 in a loan secured through a mortgage of the complete undertaking.

Moving aside from the London Extension as a whole, in terms of the Chilterns, this area was part of the southern division from Rugby to Quainton Road which incorporated the little metropolis portion to Marylebone from Canfield Place Junction. Reaching a total of forty miles, the line was divided into four contracts with Thomas Oliver and Co of Horsham covering from Rugby as far as Woodford (15 Miles 69 Chains), Walter Scott and Co of London and Newcastle on Tyne covering the next two sections to Brackley (12 Miles 39 Chains) and Quainton (12 Miles 69 Chains) while the works at Marylebone were carried out by J.T. Firbank Ltd. From a constructional viewpoint, the line which employed the services of some 5000 navvies, 150 heavy horses, 50 locomotives and 1700 wagons incorporated some unique features in that no gradient was steeper than 1 in 176, there were no level crossings, it featured the Berne continental loading gauge and all the country stations were designed on the island platform basis for reasons of economy and with the possibility of quadruple tracking in mind. As can be observed from the table below several huge engineering works were required to meet these exacting specifications:

Location	**Description of Works/Facilities Provided.**
Rugby Central	Large station and goods yards. 500,000 cubic yard embankment.
Braunston and Willoughby	Small station and basic goods yard. Viaduct of thirteen arches at Willoughby. Viaduct of nine arches at Staverton. Viaduct of twelve arches at Catesby. Catesby Tunnel 2997 Yards long from which 290,000 cubic yards of spoil was extracted and which required 30 million bricks for its lining.
Charwelton	Small station and goods yard with several extra sidings serving a local iron works. Charwelton Water Troughs 430,000 cubic yard embankment.
Woodford Halse	Two massive marshalling yards and a depot for Locomotives as well as a large repair shed for Carriages and Wagons. Station and Goods Yard, alongside which a colony of houses for employees was built.
Woodford North	Junction with the Byfield to Stratford Upon Avon and Broom Line.
Woodford South	Junction with the Byfield to Stratford Upon Avon and Broom Line.
Culworth Junction	Junction for Banbury.
Culworth	Small station with goods yard. 486,000 cubic yard embankment.
Helmdon	Small station with goods yard. Viaduct of nine arches over the River Tove and the London North Western Railway. 336,000 cubic yard cutting.
Brackley	Large station with goods yard. Viaduct of twenty two arches over the River Ouse flood plain and source. 320ft long. 180,000 cubic yard cutting.
Finmere	Small station with goods yard. 200,000 cubic yard embankment. 225,000 cubic yard cutting at Mixbury. 225,000 cubic yard cutting at Chetwode. 225,000 cubic yard cutting at Steeple Claydon.

Calvert	Small station with goods yard.
Upper South Farm	Divergence of the Manchester, Sheffield and Lincolnshire Railway London Extension from the Metropolitan Railway. New Signalbox.
Neasden	New depot for the Manchester, Sheffield and Lincolnshire Railway which would be capable of handling locomotives, carriages and many wagons.
Canfield Place	Junction from the Metropolitan for the new terminus at Marylebone. Hampstead Tunnel. (691 Yards) St Johns Wood Tunnels. (1276 Yards) Lords Tunnel. (236 Yards)
Marylebone	Terminus station with four platforms with a hotel, railway offices, carriage sheds, and a goods depot with its own warehouse.

Gaining access to Marylebone was indeed very difficult as several preparatory works had to be carried out before construction could begin, these being the clearing of nearly all the property that lay above the line's course including Lord's Cricket Ground, the demolition of Harewood Square and the building of new blocks to accommodate people who had been displaced. Naturally such large upheaval brought huge claims for compensation from Landlords, but the main inconvenience for the contractors was the restrictions put on their activities. No tunnelling or site clearance was able to take place on Sunday and the local streets were out of bounds for carts transporting spoil. Instead the materials excavated by the steam navvies and brawny labourers were loaded onto a fleet of tipper wagons which were hauled along a narrow gauge track to Canfield Place Junction where their contents were transferred to a series of Metropolitan trains heading for the area at Neasden set aside for a Locomotive Shed, Carriage and Wagon Shed, Repair Sheds, Tank House, Electricity Station, Sand Drying Plant, Usual Offices and staff colony of 150 dwellings.

Returning to Canfield Place Junction from here the line made a gradual descent before finally disappearing into a tunnel as it passed under Hampstead Baths which curved round numerous times until it emerged at Hilgrove Road. At this point the new London Extension had to cross four railway tracks and a station which belonged to the London North Western Railway, entailing partial remaking of Loudoun Road bridge, while the station of the same name required extensions to the western ends of it's platforms so that a corresponding length at the eastern end could be cut away to allow the erection of a 118ft girder bridge. Once this was complete, Firbank and Company were able to concentrate on the St.John's Wood Road tunnel which was constructed by making use of both cut and cover and tunnelling techniques. Twin bores were established throughout the whole length of the tunnel but in the end tracks were only laid through one of them. Emerging from the depths of the tunnel a mass of tracks diverged in all directions.

The Marylebone complex was laid out very neatly with a massive coal yard in the south western corner bounded by Great Carlisle, North Street and Grove Road. Further down Grove Road shortly after the thoroughfare turned into Lisson Grove, at the point where it effected a junction with Rossmore Road, there was built a wide entrance which led into a capacious goods yard, equipped with a superb five storey warehouse with hydraulic and electric machinery powered by independent plants, a giant crane, several sidings and numerous goods offices. On the other side of the main lines, the contractors erected a long carriage shed with scope for extensions, an oil gas works for producing fuel for the lighting, a complex array of sidings and a wharf. The latter all lay north of Rossmore Road which effectively created a division between the Goods/Depot and Passenger areas.

South of Rossmore Road bridge, after the small station throat had been negotiated, the contractors established four twelve carriage length platforms whose length was partly covered by a triple span canopy and a temporary awning which overhung platform four. Five tracks with hydraulic buffer stops at the end, were laid into the complex, one being a long siding running beside that leading into Platform One, while the others served Platform Two (adjoining the cab and parcels carriage way down from Rossmore Road), Three and Four respectively. Turning to the reception area of the station, Marylebone boasted a large concourse which gave a light and airy atmosphere unlike some of its counterparts and naturally it had all the usual features such as Refreshment Rooms,

Waiting Rooms, Booking Hall, an Underground Station, Kiosks and Toilets. On the front, the building were faced with "Redbank" Pressed Bricks with dressings of buff terra cotta similar to that used on the massive hotel which was constructed shortly after between the station and the Marylebone Road.

Eventually by the middle of July 1898 the finishing touches had been put to Marylebone, such that by the 25th the old Manchester, Sheffield and Lincolnshire Railway, whose name had been changed to the Great Central Railway in anticipation of a grand opening of the London Extension, started running coal trains to compact and consolidate the new trackbed. Five days afterwards the Great Central sent a message to the Chairman of the Metropolitan that under an agreement drawn up by his predecessor Sir Edward Watkin, their trains could run off onto the Great Western Railway branch from Aylesbury to Princes Risborough. However John Bell the new Chairman had other ideas since he detested William Pollitt of the Great Central whose apprentice he had once been and he knew that with Sir Edward Watkin having been retired from railway matters some four years previously, he could cause a bit of a stir over what he regarded as a rather unfriendly act. In an amazing move Bell went off one morning with a band of volunteers to Quainton Road in order to place the signals at danger personally and thus block the passage of the first train. As it drew to a shuddering halt at approximately 3.00am in the morning, the locomotive crew were astounded to hear from Bell that the train would not be able to proceed until some resolution could be worked out with William Pollitt. For the footplatemen, no choice existed other than to make arrangements for them to be towed back from where they had come.

After this episode, made all the more surprising by the fact that the Metropolitan had invested heavily in upgrading and doubling the line from Aylesbury to Verney Junction in readiness for the coming of the London Extension, the Great Central Railway rashly decided to sever all links with the Metropolitan which would not only seriously affect the latter's traffic but also cripple the Great Central financially as their only option to get traffic in and out of Marylebone was by building a new line from Canfield Place to Grendon Underwood. To this end the Great Central after meeting with the Great Western managed to form an alliance, on the grounds that the former wanted to create an avoiding route, while the latter were seeking a more direct line to Birmingham, which could make use of the same course, so a bill was duly put before Parliament in 1899. Meanwhile around the same time on the 9th March, the concourse at Marylebone had been laid for a huge banquet which was to precede a small ceremony where the Rt Hon. C.T. Ritchie, MP, President of the Board of Trade, pulled a small silver lever attached to locomotive 861 to set it off, thus marking the opening of the London Extension to passenger services.

Within two years, under the direction of the Resident Engineer R.C. Sikes, construction was begun by Messrs Thomas Oliver and Sons of Rugby on the Great Central section from Neasden as far as Northolt where it met the Great Western route from Royal Oak Common through North Acton, Park Royal, Perivale and Greenford whose construction was entrusted to Messrs Pauling and Company of Westminster. Their miraculous efforts, ensured that enough progress had been made on the 15th June 1903 to enable several passenger trains to be run right up to Park Royal for the Royal Agricultural Show. Meanwhile another firm Messrs Pattinson and Son of Westminster hastened along with the construction of the stations, at Wembley Hill, Sudbury Hill and Sudbury and Harrow Road, all of which were very lavish affairs boasting both fast and slow main lines in each direction, an impressive array of buildings and large goods yards. Undoubtedly such extravagance obviously accounted for part of the increase in building costs, but the majority of the extra outlay stemmed from a problematic engineering feature at Wembley. At this point a tunnel had been originally chosen, but rather unwisely the engineer felt that a massive cutting would be better which entailed the extraction of some 640,000 cubic yards of spoil, the digging of a 30 foot foundation and the building of a wall 158 feet long by 30 feet high.

From the point where the lines from Marylebone and Paddington met at Northolt, the Great Western and Great Central companies decided on a short length of quadruple track as far as Ruislip. Midway along this section the railway passed over the Ealing to Ruislip line on a rather novel design of girder bridge. A little way along from here the railway crossed the Metropolitan's Uxbridge route, just prior to reaching Ruislip station on the other side of which two mile long water troughs were laid in the track for the non stop pick up of water.

Approaching Denham, Messrs Pauling and Company the contractors had to locate the station with it's two main lines, two loops, well appointed platforms and adequate goods yard on a massive long embankment which ran up to the Grand Union Canal where it was found necessary to erect a splendid viaduct of nine arches before continuing on a further embankment to where a four arch viaduct was constructed to cross the River Colne. Viaducts were a common feature on this stretch and before Gerrards Cross had been reached another two of five spans each bridged the River Misbourne and the Chalfont Road respectively. A short distance down the line saw the start of the rolling Chiltern Hills where embankments gave way to the need for deep cuttings, running for one to one and a half miles ranging in depth from forty five to eighty five feet, one of which dwarfed the rather lavish station at Gerrards Cross. From here the line briefly sat proudly on an embankment before entering a series of deep cuttings once again, whose construction entailed the destruction of several beautiful trees in the Wilton Park Estate, which ended near the site of the new Beaconsfield Station where an identical facility to Gerrards Cross was constructed.

Beyond Beaconsfield, three really spectacular engineering feats were created, these being an eighty foot high embankment, a long tunnel at White House Farm where unfortunately six men perished during its construction and a structure of three spans known as Penn Viaduct which at the request of local landowner Sir Philip Rose had to be decorated externally with ornamental red bricks. Heading on from what soon gained the nickname Sir Philip Rose's viaduct, for some distance the line clung to the hillside above the Beaconsfield Road until it reached Loudwater.

As the new Great Western and Great Central joint line approached the outskirts of High Wycombe, it joined up with the diminutive Wycombe Railway from Maidenhead which followed a parallel course on the other side of the Wye Valley before sharply cutting across the Marshes and running into the town's station. The very nature of the hilly terrain and chalky soil as the navvies who built the Wycombe Railway found, proved it to be a difficult area to enter and emerge from, so now with widening in prospect the work would be even harder, especially if disruption in the local community was to be kept to a minimum. Of course numerous cuttings had to be have extra spoil carefully cut out of them, which was taken to widen the embankments. Additionally, all of the bridges which were originally built with single spans, required reconstruction in order to accommodate the double track.

Double track already existed at the station where the platforms faced each other, although the needs of both the Great Western and Great Central Railway companies dictated that there should be four tracks located here to cater for stopping trains while the other two would allow through movements. However space was at a premium such that in order to accommodate the extra rails, the Up side platform had to be demolished and relocated a whole length further south of its original position with a new subway link to it's down side counterpart which was bestowed with a new set of facilities. Since the old Up side platform had been built into the side of the hill a large retaining wall had been erected behind it, which needed to be demolished to allow the navvies to carve out a further slice of the hill, after which an incredible sized replacement was erected using up one and a quarter million Staffordshire Blue bricks.

Leaving High Wycombe another contractor Messrs Mackay and Davies became responsible for the works as far as Princes Risborough and Haddenham. On this section the first major engineering feat came at Frogmoor where a second more sturdy viaduct had to be provided for express trains alongside the existing bridge. Past these two structures sufficient spoil was dumped to make an extra area for new sidings which would be capable of accommodating 200 wagons. Moving on towards West Wycombe the engineering works became much lighter though at the Station here again the Up side platform had to be rebuilt to allow some room for the through roads but in the event these were never laid. Meanwhile on the Down platform side a completely new set of buildings were erected.

Saunderton the next port of call, having opened on the 1st July 1901, ironically not only found itself having to be remodelled in a conventional fashion, but also it gained the reputation of being the start for a rather unusual feature where the Up line parted company from the Down line. The single track of the old Wycombe Railway became the new Down line which was on a falling gradient. To the east of the latter, much heavy excavation took place to produce a cutting and tunnel so that the up line would not have to climb up the gradient which began on leaving Princes Risborough.

At the latter location significant changes occurred to the old layout necessitating the construction of a totally new station further south of the original which offered two through lines with two loops serving a lavish up and down platform with bays for the Aylesbury and Watlington and Thame Branches. For goods traffic from these and the main line the yards at both ends of the station were given an increased wagon holding capacity.

On its exit from Princes Risborough the line passed the superb new north Signalbox with its lever frame the length of a cricket pitch, before heading over four miles on a straight course until it encountered the five span Chearsley viaduct. A mile and a half further along following the usual pattern came Haddenham station which was the final port of call prior to the line reaching the site of the proposed Ashendon Junction some four miles away. At this point the Great Central assumed sole ownership of the line as far as Grendon Underwood where the main London Extension was regained via a flat junction. On this lightly engineered stretch two intermediate stations were established at Wotton and Akeman Street.

Returning to the Haddenham/Great Central stretch which L.P. Knott and Company had been responsible for constructing, at Ashendon approximately half way along the line, the Great Western Railway commissioned Walter Scott and Middleton to provide a new flying junction to gain access to the Aynho Junction extension which was begun in 1905 around the same time as the joint line opened for goods traffic. Five months later on the 2nd April 1906 the first passenger services began to run from Marylebone and Paddington, an event which was marked by a public luncheon at the town hall in High Wycombe hosted by the local Chamber of Commerce for some of the important figures connected with the line. In the middle of the event the Mayor offered a toast "Success to the new line" which brought an immediate response from Alex Henderson of the Great Central who stated that on the one hand the Great Central would offer access to an "Inexhaustible store of the Midlands", while the Great Western allowed the local community an "Instant Gateway to the West of England". Meanwhile the horizons of the line were being further expanded with rapid progress occurring on the Ashendon Junction to Aynho Junction line which required the construction of the following features:

Ashendon	Flying Junction where the up line of the Great Western passed over the two tracks of the Great Central. Signalbox. Crossing of the Metropolitan Railway's Brill Branch. 191 yard tunnel.
Brill & Ludgershall	Station with two through roads, two loops through platforms containing ornate brick buildings, a footbridge and a goods yard.
Blackthorn	Station with two through roads, two loops through platforms containing ornate brick buildings, footbridge and goods yard. Embankments of large proportions for three miles up to Bicester. Crossing of the London North Western line from Oxford to Cambridge.
Bicester	Station with two through roads, two loops through platforms containing ornate brick buildings, footbridge and goods yard. Embankment with a 1 in 200 climb. Large cutting up to Ardley.
Ardley	Station with two through roads, two loops through platforms containing ornate brick buildings, footbridge and goods yard. Cutting. Tunnel 1,147 yards long. Souldern Viaducts 580 yards and 400 yards.
Aynho	Conventional two platform Station for the Park akin to a halt.
Aynho	Flying Junction with the Oxford to Didcot to Paddington route. Signalbox.

Leaving the newly created junction at Aynho, the Great Western Railway expresses followed the existing 1852 route all the way to Birmingham via Leamington Spa and Banbury where a brand new locomotive depot and sidings were established in 1908 just two years before the remainder of the cut off line opened to both goods and passengers on the 4th April and 1st July 1910 respectively.

Marylebone Station concourse shortly after opening with the main entrance over to the left under the arch, while just to the right lie the 1st and 3rd Class Ladies Rooms and the Booking Office. Note the all glass canopy which was designed in such a way so as to give the whole area a light airy atmosphere. (Leicester Museums)

After a sumptuous banquet on the concourse at Marylebone all the guests assembled around a cordonned off area to watch the Right Honourable C.T. Ritchie President of the Board of Trade pull the lever on the dais and set off the first train on the London Extension. (Leicester Museums)

A superb view of a Great Central Railway A5 class tank engine No 450 seen gently easing away from the throat of Marylebone station past the carriage sidings with a down Aylesbury train on the 12th May 1915. Note the Robinson anti collision buffers on the carriages, a feature which had only recently been introduced when this photograph was taken. (D.Edwards)

On the 13th March 1947 a London North Eastern Railway B1 Class No 1123 is seen climbing away from Platform No 1 at Marylebone with a rake of empty carriaging stock bound for Neasden Depot. (Authors Collection)

Sometime in 1938 London North Eastern Railway B17 No 2854 Sunderland is seen at the head of the 10am Marylebone to Bradford express. Sadly due to repairs the locomotive was indisposed during the previous year when Sunderland were playing in the Cup Final, prompting the LNER to transfer the name and number to a substitute engine to celebrate the occasion. (Authors Collection)

An overall view of Marylebone Station on the 30th April 1957, showing the Great Central Hotel in the background, the train shed and the signal cabin on the platform and numerous parcels vans in evidence. To the left of these is a local train for Woodford Halse. Meanwhile over on the right is a "Master Cutler" working headed by British Railway's Class 5MT 4-6-0 No 73157. (Courtesy M.Bromley)

On a Sunny 16th June 1948 a Black Five class 5mt No 45253 breathes a sigh of relief after the long haul as she eases her express from Manchester London Road into Platform One of Marylebone station. (B.Brooksbunk)

Amongst a pack of shiny Diesel Multiple Units one of the old order ex London Midland Scottish Railway Black Five Class No 44858 in a deplorable condition makes the climb away from Marylebone with the last working of the day to Nottingham at 4.38pm on the 15th May 1966. (K.C.H. Fairey)

Great Central 8B Class 4-4-2 No 269 all decked out for the haulage of a Royal Train by the looks of it has just taken on a full load of coal and is now preparing to go and find her train in Neasden carriage sidings. (Authors Collection)

The date is the 18th August 1921 and Sam Fay Class No 423 Sir Sam Fay is resting outside Neasden Shed along with a whole host of engines some possibly of the same class. (Authors Collection)

On the 20th May 1933 an ex Great Central Railway 9P class No 6167 (originally No 1167) Lloyd George whose name was removed stands in light steam as she waits for the signal to proceed. (Authors Collection)

All varieties of ex Great Central and London North Eastern Railway locomotives could be found in residence at Neasden shed and here is a rather inactive F2 No 5773. (Authors Collection)

A Great Central Railway Class 9P 4-Cylinder 4-6-0 No 1167 LLoyd George is seen on a Manchester express near Wembley. Note the seven varnished teak carriages with their raised numbers, anti telescoping devices and carriage boards. (Authors Collection)

BAKER STREET TO HARROW

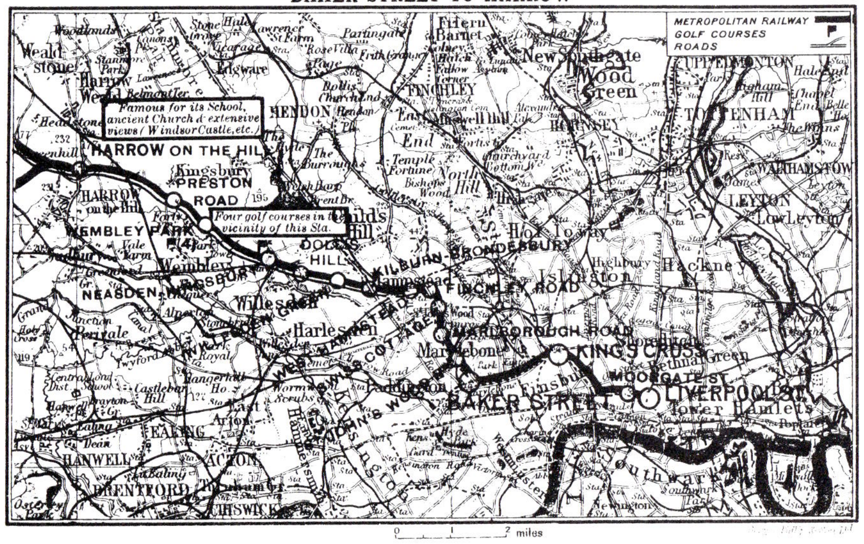

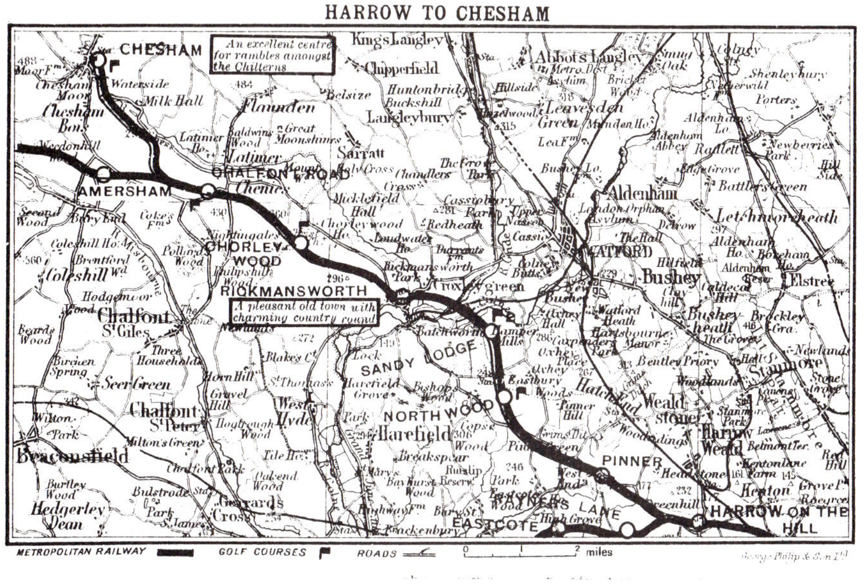

The old version of Harrow on the Hill station was really much more attractive than the new one as can be seen from this 1906 view where the station is remarkably quiet which is probably why the staff on the right under the platform sign have time to pose for the photograph. (Lens of Sutton)

The modern version of Harrow on the Hill station was begun in 1939 and saw the entire removal of the buildings on the left at road level and those on the platforms. Afterwards the latter were lengthened and provided with several ugly concrete notice panels some of which were adorned with a double lamp standard while under the no frills canopy lay kiosks, a staff office and a light airy waiting room. At road level the buildings formed a bridge which contained a booking hall, a relay room, signal box and offices. (Lens of Sutton)

The date is the 10th September 1946 and a rather grubby locomotive No 2916 is seen sitting at Moor Park station with a Marylebone to Woodford local train. (H.C.Casserley)

An interesting photograph of Moor Park in 1961 showing the remains of the original station and the main lines through it. Soon from the chaos will emerge an identical platform and track layout to that on the left into which the sleek Metropolitan Electric Locomotive No 4 is seen drawing an Aylesbury to Liverpool Street train. (H.C.Casserley)

Until the early 1960's Metropolitan trains had to exchange traction at Rickmansworth which was a very time consuming exercise. Here a London North Eastern Railway engine 6150 is waiting to hand over to an electric locomotive to take the train on to Baker Street. (H.C.Casserley)

A splendid period picture of Metropolitan Class H Tank engine No 105 steaming away quietly at the north end of the sharply curved Rickmansworth Station waits to depart with a stopping train bound for Verney Junction on the 22nd April 1933. (H.C.Casserley)

With a Baker Street to Aylesbury train, an ex Metropolitan Class H locomotive is seen climbing up the gradient north of Rickmansworth in April 1938 approximately a year after the locomotive was purchased by the London North Eastern Railway. (Authors Collection)

Bunker first a Metropolitan K Class engine No 115 regains the main line to head south to Chorley Wood after picking up a few wagons from the yard at Chalfont and Latimer on 22nd April 1933. (H.C.Casserley)

A former London Midland Scottish Railway Ivatt designed 2-6-2T locomotive No 41272 edges away from Chalfont and Latimer Station at the back of a Chesham Shuttle on the 19th May 1959. Until the 1960's it was standard practice for British Railways to provide an engine for this duty while the Metropolitan would supply the carriages. (B.Brooksbunk)

Ex London North Eastern Railway Class A3 4-6-2 No 60054 Prince of Wales makes a fine sight as she thunders along with a well loaded up South Yorkshireman on 20th August 1953 past Boughton and Sons' works just outside Chalfont and Latimer Station. (Courtesy M.Bromley)

As the evening shadows lengthen on the 26th July 1947, a London North Eastern Railway tank engine No 7420 simmers gently in the platform at Chesham whilst waiting to head off to Chalfont and Latimer. (H.C.Casserley)

14D Neasden based Black Five 5mt class No 45215 thunders along through Amersham station with a well loaded 13.30pm Marylebone to Wakefield express on the 19th May 1959. Not long after this scene was transformed as an extra platform was added to the station on the far right at the same time as London Transport electrified the tracks. (B.Brooksbunk)

On the 9th September 1961, the last day of Metropolitan train working to and from Aylesbury Town and the end of steam hauled trains on the route, an immaculate Fairburn 2-6-4T No 42070 hauls a farewell special into Amersham. The engine was built in Brighton in 1950 and withdrawn only fifteen years later. (J.A.Fleming)

In this divine view taken in the 1920's Metropolitan "H" Class tank engine No 106 is seen heading bunker first on a London Baker Street to Aylesbury working just south of Great Missenden (Authors Collection)

Looking down over Great Missenden station from what later became Trafford Close a Verney Junction bound train sits in the down platform waiting to depart. Note the horsebox and two parcels vans and the down platform buildings which were demolished in the 1960's. (Lens of Sutton)

A splendid period view of Wendover Station where a Great Central 11B class No 1038 has stopped to pick up a large number of passengers bound for Marylebone in 1905. (Lens of Sutton)

Stoke Mandeville

During the final year of Metropolitan operations beyond Amersham ex London Midland Scottish Railway Fairburn 4MT Class 2-6-4T draws into Stoke Mandeville with a train of six Dreadnought carriages forming an Aylesbury to Baker Street stopping service. (Authors Collection)

During a quiet moment on a hot summer's day in June 1960 between trains from Aylesbury's No 4 Platform can be seen the neatly compact shed, watering column and coaling stage beside which is sitting Derby Built Class 2MT No 84029. (Authors Collection)

During an overcast day in July 1961, former London North Eastern K3 No 61834 negotiates the pointwork by Aylesbury South Signal cabin with a Manchester London Road to London Marylebone express. (A.Vaughan)

An absolutely magnificent shot of a Metropolitan K class 2-6-4T engine No 113 at the head of a goods train is seen waiting for a signal to proceed southwards from Aylesbury Station's No 2 Platform. (Authors Collection)

On the 15th March 1930 having run round its train No 222 a Great Western Railway workhorse is seen simmering gently in PLatform No at Aylesbury waiting to depart bunker first for Paddington (via Maidenhead). In the background is the shed and note the ex Brill Tramway carriage body near the bunker of the engine on the ground. (H.C.Casserley)

London North Eastern Railway F1 Class No 5594 has just arrived at Aylesbury station's No 2 Platform with what was the 3.30pm departure from Verney Junction. Note the signalbox and goods yard in the background which sadly have long since disappeared. (Lens of Sutton)

Aylesbury station during a Summer afternoon in the 1920's with a Leicester Central to Marylebone train headed by an ex Great Central 4-4-0 D10 Class engine No 5438 which is seen receiving a bit of attention to its wheels prior to leaving Platform No 2. Alongside it in Platform No 1, one of the immense Metropolitan Class H engines waits quietly with an all stations to Baker Street working. (Lens of Sutton)

Whilst a London North Eastern Railway J11 "Pom Pom" class locomotive No 5201 sits on one of the shed roads taking a much needed drink, the one coach motor train headed by F7 class No 8307 that used to push and pull its way along to Verney Junction and back is seen waiting for custom in platform 4 on the 8th April 1933. (H.C.Casserley)

An ex London North Eastern Railway N5 0-6-2T No 69273 is seen heading with it's two carriages off the single line from Princes past Aylesbury South Signalbox towards the station. (Courtesy M.Bromley)

The remains of Waddesdon Manor station looking northwards in the late 1950's with the buildings and goods yard long since gone. (H.C.Casserley)

Some three years after closure there is little evidence remaining of the former Waddesdon Manor station which an ex Metropolitan K Class locomotive is seen passing with a heavy coal train on the 17th June 1939. (H.C.Casserley)

A delightful view of the ex Great Eastern Railway 2-4-2T No 8307 in railmotor formation at Quainton Road in early 1936. Up until 1935 Quainton Road had been the junction for the Brill Tramway. (H.C.Casserley)

One summer day in 1934 the penultimate year of operation, ex Metropolitan A Class 0-4-4 tank engine No 23 complete with condensing apparatus for tunnel working is seen with its eight wheeled coach waiting to depart for Brill. (Authors Collection)

A splendid view of Westcott Station from the roadside, showing the employees cottages and the level crossing in the background, while in the foreground are the station buildings and entrance to the goods yard. (London Transport)

An overall view of the tramway station at Wotton in 1926 showing the substantial goods transfer shed and siding in the foreground while in the distance is the small platform behind which lay another freight handling siding. (Authors Collection)

Wood Siding which was a little station tucked away in the trees towards the end of the Brill Branch which like its counterparts on the route had the proud boast of being a part of the larger more sophisticated London Underground system, until it closed in 1935. (Authors Collection)

Although this picture has appeared a few times before in books it is one which is somehow always captivating. The wooden building on the left with its tell tale signs from the removal of the noticeboards and the rather resigned look on the drivers face give an all too clear indication that this is the end for the little Brill Branch as do the crowds of people and the bowler hatted official standing near the guards van. (London Transport)

Quite a distance from its old stomping ground but enjoying retirement in comfortable surroundings is Aveling & Porter No 2 (once a Brill Tramway locomotive) which was rescued and lovingly restored for display at the London Transport Museum. (Author)

Having just left Quainton Road station which can be seen in the background, an ex London North Eastern Railway V2 2-6-2 No 60961 makes light work of her six carriage load forming a Marylebone to Woodford Halse local working. On the left are the disused Metropolitan lines which led to Verney Junction. (Authors Collection)

On the 2nd May 1936 tiny ex Great Eastern Railway 2-4-2T No 8307 in railmotor formation is seen sitting at Winslow Road with a Verney Junction train. (H.C.Casserley)

An interesting overall view of Calvert Station which was the first station going north on the Great Central London Extension, with 5.00pm Marylebone to Woodford Halse train just drawing in on the 9th May 1959. Note the wagons on the right which were used for carrying large quantities of bricks from the nearby works. (Authors Collection)

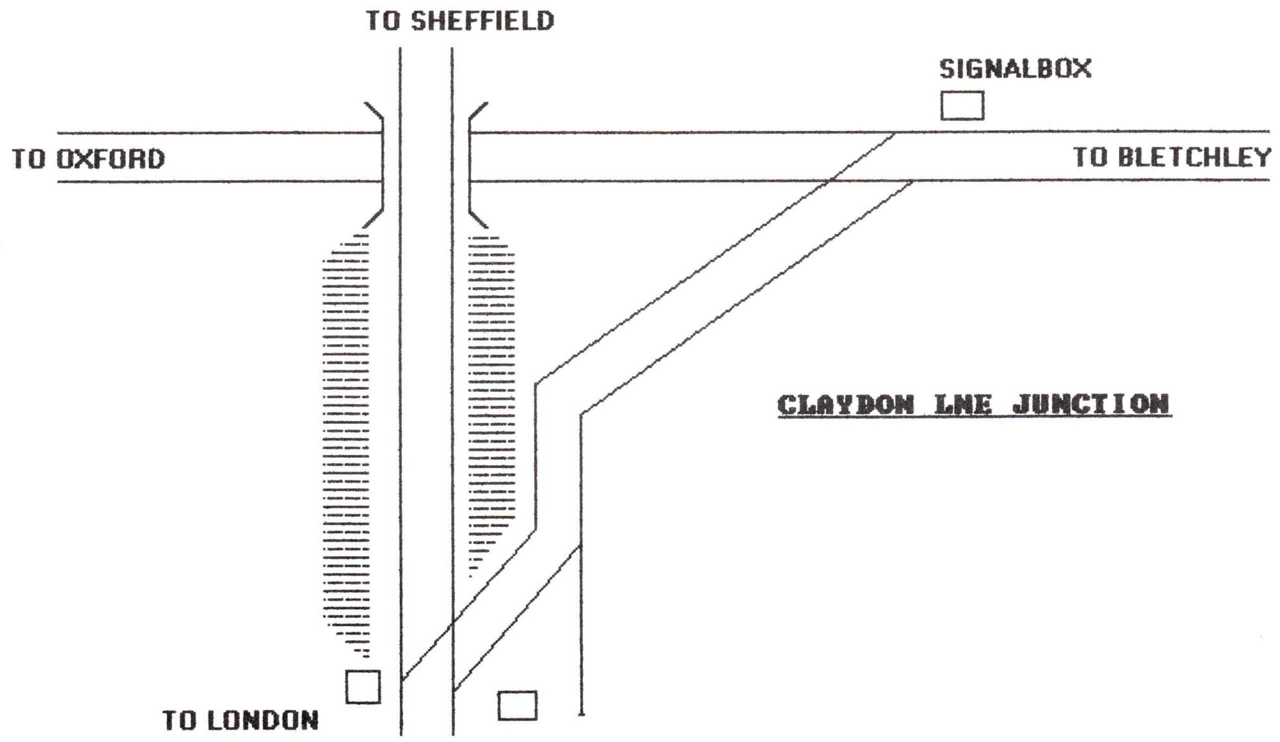

Verney Junction on the 27th February 1960 looking towards Bletchley with a train waiting to depart from the platform in the background while in the foreground is the station's ornate garden and former Metropolitan line platform. (Courtesy M.Bromley)

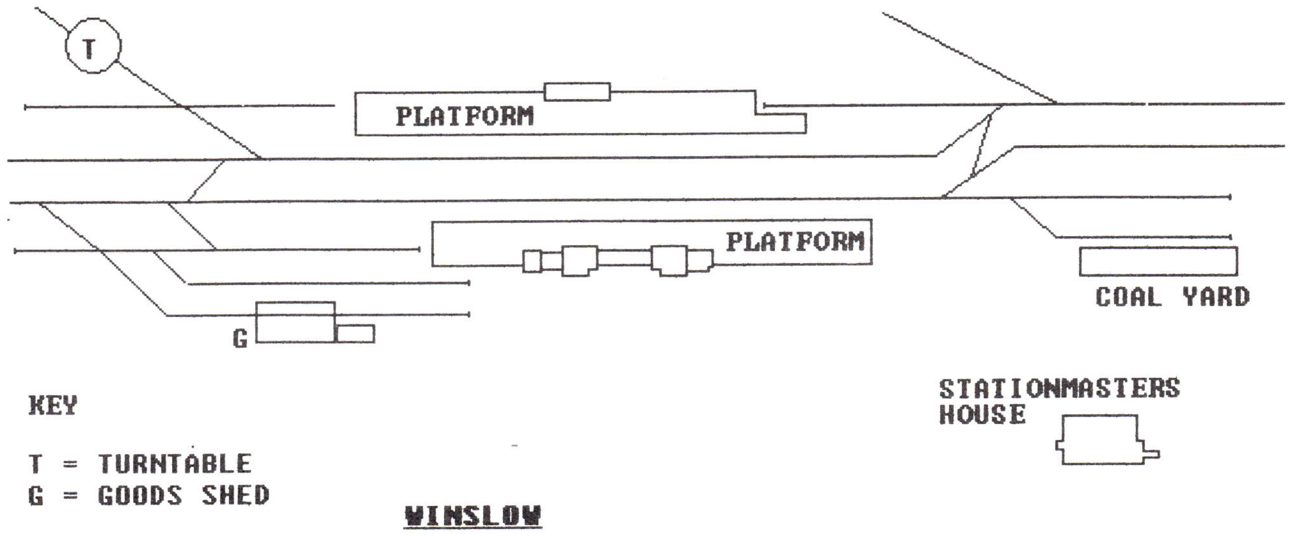

Having arrived at Wembley with a packed football special from Manchester for the Cup Final on the 7th May 1960, ex London North Eastern Railway No 61169 is seen backing her train past Neasden South Junction Signalbox on the goods reception line. (Authors Collection)

With a long eleven carriage rake behind, Ex London North Eastern Railway Class A3 4-6-2 No 60052 Prince Palatine heads north with the Master Cutler which left Marylebone at 18.15pm and arrived at Sheffield some three and three quarters of and hour later. (Courtesy M.Bromley)

The rather large Signalbox at Sudbury Hill Harrow in 1967. (Authors Collection)

Leaning out of the front window of the first carriage of a Marylebone to Mansfield special on the 16th July 1959, the photographer has managed to capture a splendid shot of the elegant up side station building, track layout and the ex London North Eastern Railway B1 Locomotive No 61140 which is in charge of the train. (Courtesy M.Bromley)

An example of a somewhat functional yet simple station is seen here in the form of Northolt Park which opened on the 1st March 1907 to serve a nearby housing development. (Authors Collection)

Ruislip Gardens as seen here was a typical example of a Great Western Railway basic station crudely constructed from planks of wood or sometimes old sleepers. A shelter was usually provided on each platform in a Pagoda style corrugated iron hut. When the Central line extension of the underground arrived a new station with the same name was opened for the tube trains and the old Great Western facility closed. (Lens of Sutton)

It is the 4th May 1957 and there is great excitement on board a Cup Final Excursion as ex Great Western Railway 4-6-0 Hall 4900 Class No 5010 Restormel Castle charges through South Ruislip to Wembley. (H.C.Casserley)

As the shadows of the evening sun lengthen over the Great Western and Great Central Joint Line, an ex Great Central L1 2-6-4T No 5339 dashes along on an up mixed goods near West Ruislip on the 3rd May 1951. (Real Photographs)

West Ruislip Station from as seen from the down platform showing the footbridge end of the up platform and the two fast lines on the up side of which can be seen an express bound for Paddington which had probably originated from Wolverhampton. (Lens of Sutton)

Having belonged to two previous owners one of the large ex Great Central Railway 4-6-2T A5 Class Locomotives No 69801 sidles effortlessly over Ruislip Water Troughs on a Woodford Halse to Marylebone train. (Courtesy M.Bromley)

A Wolverhampton to Paddington express hauled by ex Great Western Railway Castle Class No 5010 Restormel Castle is seen passing Denham Golf Club Halt in June 1956. (Courtesy D.Edwards)

Dominating the scene south of Denham is a fine example of Great Western lower quadrant signals under which a ex War Department 2-8-0 No 90208 is about to pass with a heavy mixed goods train. At rear of the train on the down fast road a 6100 class waits for a passage as she lets a local passenger train head away from Denham Station. (Adrian Vaughan)

Rather effortlessly one of the Thompson rebuilds of an old Great Central design in the shape of an 01 Class 2-8-0 No 63736 thunders along through the station at Denham on the fast line with a northbound freight. (Adrian Vaughan)

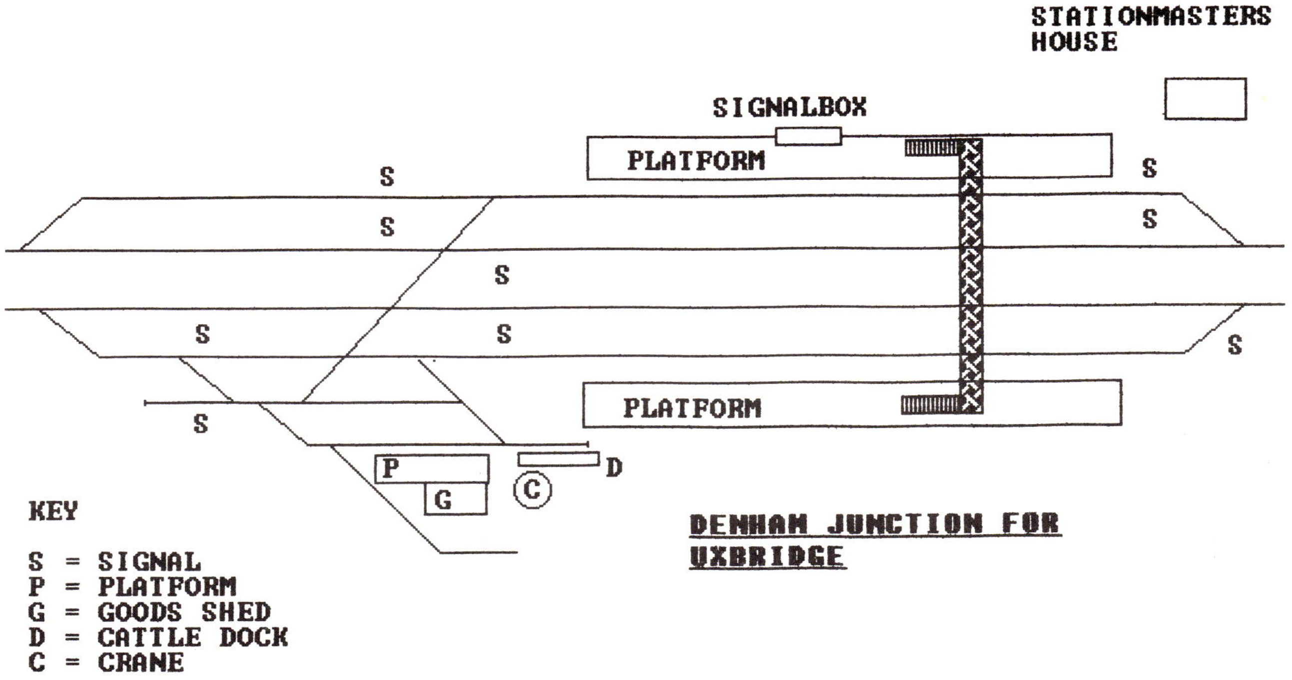

Ex London North Eastern Railway Gresley built A3 Pacific 4-6-2 No 60052 Prince Palatine thunders along near Harvil Road, Denham with a Marylebone to Manchester express on the 3rd April 1954. (D.Edwards)

Having left Marylebone some while ago an ex Great Central Railway A5 class tank engine No 69805 is seen making its way with a packed six coach train towards Gerrards Cross while on its way to Woodford Halse. (Adrian Vaughan)

A Birkenhead to London Paddington express hauled by King class 4-6-0 No 6020 King Henry IV rumbles through Gerrards Cross station on the 1st July 1957. This engine was built in 1930 and fitted with a double chimney in 1956 six years before she was withdrawn and scrap at Cox and Danks' yard at Park Royal. (J.A.Fleming)

On the 7th July 1956 a 1E Bletchley based Black Five 5mt Class 4-6-0 Mixed Traffic engine No 45307 is seen about to trundle along with an up parcels train through one of the centre roads at Gerrards Cross. (J.A.Fleming)

During late Summer 1947 a former Great Central Railway B8 class 4-6-0 No 5004 sidles away from Gerrards Cross with an eight coach Marylebone to Woodford Halse local train. (Authors Collection)

On a clear March afternoon just four days into the Spring of 1950, an ex Great Western Railway King Class 4-6-0 No 6008 King James II heads along with ease near Seer Green with the 16.10pm Paddington to Birkenhead express. (B.Brooksbank)

Seer Green and Jordans Halt which as the notice proudly proclaims is the alighting point for Beaconsfield Golf Club on the 11th January 1961. (Lens of Sutton)

Back into the daylight once again comes ex Great Western Railway 4900 4-6-0 Hall Class engine No 4987 Brockley Hall as she emerges from White House Farm Tunnel (between High Wycombe and Beaconsfield) with a class H freight. (A.Vaughan)

Ex London North Eastern Railway L1 class 2-6-4T No 67776 ambles along gently on a High Wycombe to Marylebone local train near Seer Green in July 1953. (J.A.Fleming)

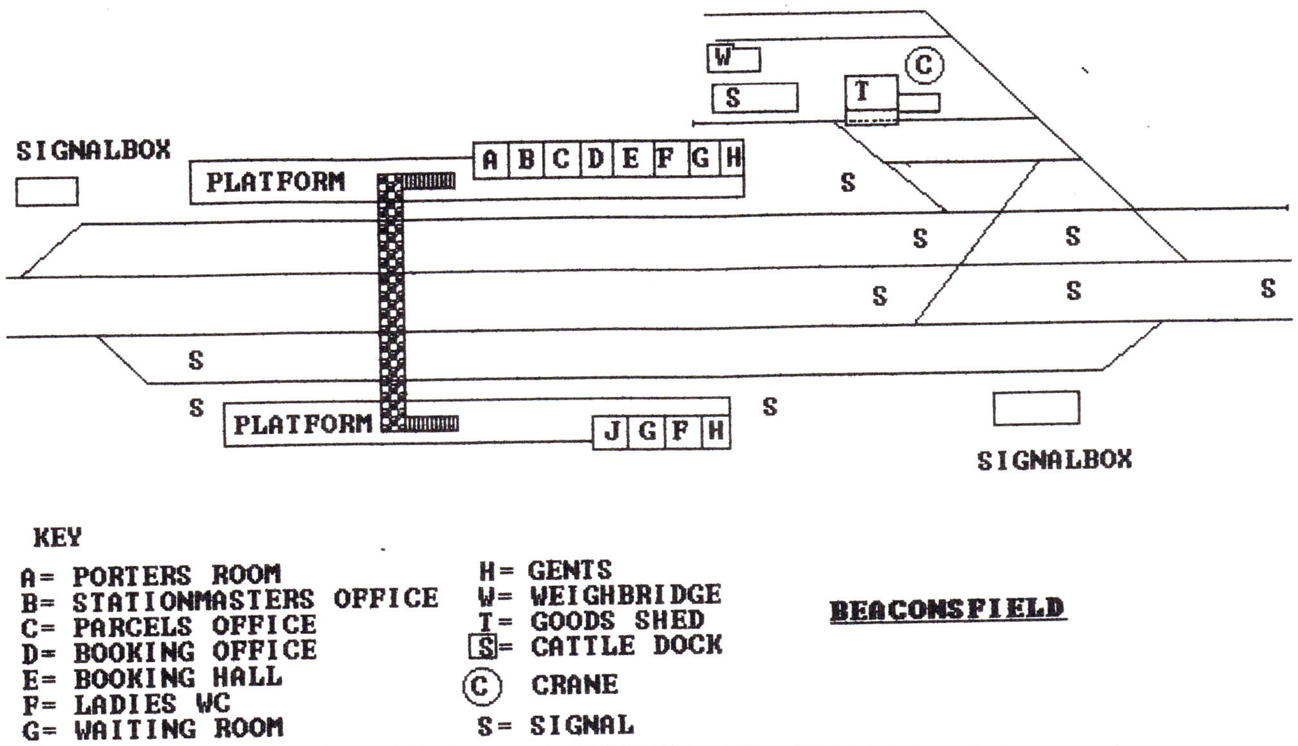

An unidentifiable Great Western tank engine simmers gently in Beaconsfield station whilst on an afternoon Paddington to Aylesbury working in 1912. (Lens of Sutton)

The station staff and passengers are ready for action as a London Paddington to Birkenhead express sidles into the down slow line to call briefly at High Wycombe during July 1958, headed by a King Class 4-6-0 No 6006 King George I, being one of a number of engines that had been fitted with a double chimney in 1956. (J.A.Fleming)

Speeding northwards away from High Wycombe station on the down fast road, the down Birmingham Pullman is dwarfed by the massive retaining wall which had to be constructed to allow widening of the line in 1905/1906. The actual train itself boasted a sumptuous interior and a handsome medium blue livery on its exterior with a wide white band around the windows. This service began running in 1960 but sadly was withdrawn on the 6th March 1967. (J.A.Fleming)

With snow still in evidence on the 9th March 1958 having set off from Aylesbury an ex Great Western Railway 1400class 0-4-2T No 1473 with a one coach auto train is seen approaching High Wycombe where it will terminate. (B.W.Leslie)

At the end of June 1957 a "2251" class 0-6-0 No 2276 is ambling merrily along from West Wycombe to High Wycombe with an up mixed goods train from Banbury Goods Yard to Old Oak Common. 2276 was built at Swindon in September 1934 and was withdrawn in September 1962, before being despatched to J.Cashmore Ltd of Newport for breaking up. (J.A.Fleming)

On the Great Western and Great Central cut off line, there is still much to be done to complete West Wycombe station seen here towards the end of 1904. Although space existed unlike other locations no centre roads were ever laid. (Leicester Museums)

WEST WYCOMBE STATION

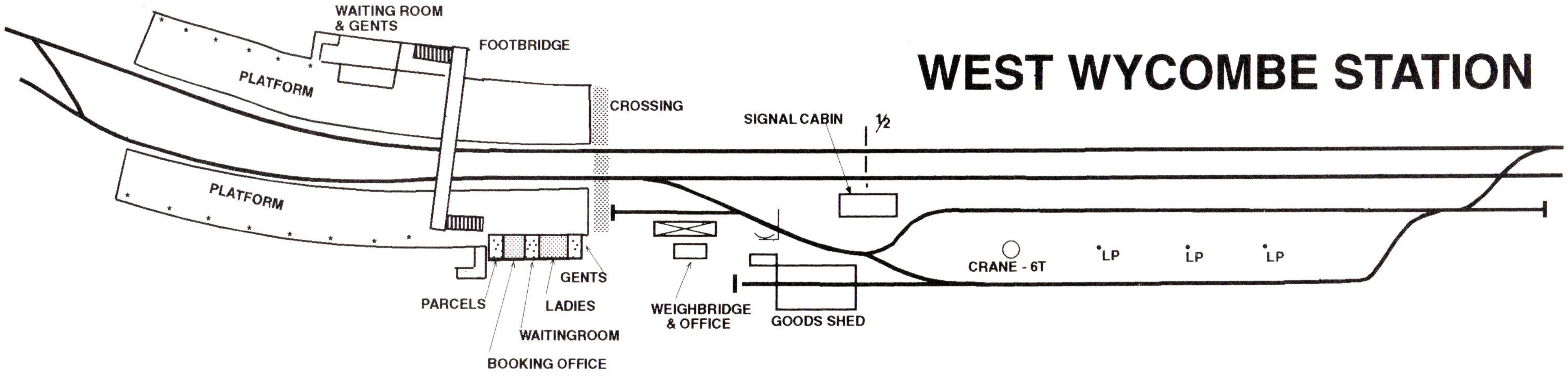

West Wycombe station in happier times with a busy looking goods yard and a down platform with relatively well kept buildings as seen from the up side. Within just two years after this picture was taken, the station sadly closed due to pressure of competition from coach services. (Lens of Sutton)

Another 84A Wolverhampton Stafford Road based engine in the form of Castle Class 4-6-0 No 5070 Sir Daniel Gooch charges along with a Wolverhampton Low Level to London Paddington express on the 17th July 1958. (J.A.Fleming)

A “2800” class 2-8-0 No 2830 allocated to 84B Oxley Shed at the helm of a heavily laden mixed goods train between Saunderton and West Wycombe on 9th July 1957. (J.A.Fleming)

In British Railways days ex Great Western Railway Castle Class 4-6-0 engine No 7024 Powis Castle based at shed 81A Old Oak Common, making light work of a Wolverhampton Low Level to London Paddington express between Saunderton and West Wycombe on 7th July 1957. (J.A.Fleming)

Another of the ubiquitous ex Great Western Railway King Class 4-6-0’s No 6014 King Henry VII from Wolverhampton Stafford Road Shed storms along between West Wycombe and Saunderton with a London Paddington to Wolverhampton Low Level express on the 14th July 1958. (J.A.Fleming)

One of those splendid ex London North Eastern Railway V2 class engines No 60911 in British Railways days passes the little used Saunderton station whilst working the 12.15pm London Marylebone to Manchester London Road express on the 27th December 1958. (B.W.Leslie)

Saunderton Station in the mid 1960's looking towards Princes Risborough during a quiet period. (Lens of Sutton)

An ex Great Western Railway King class 4-6-0 No 6011 King James I approaches Princes Risborough station on 2nd July 1956 with a London Paddington to Birkenhead express while an ex London North Eastern Railway L1 2-6-4T is waiting to head off with an all stations to Marylebone train. (J.A.Fleming)

A really splendid picture showing an overall view of the south side of Princes Risborough station with the signal cabin and busy goods yard in the background, while in the foreground on the left are the exchange sidings for both the Thame and Watlington Branches past which is running an ex Great Western Railway Castle Class 4-6-0 No 5019 "Treago Castle" whilst working a Wolverhampton to London Paddington express on the 1st July 1959. (Courtesy D.Edwards)

With very little activity in evidence apart from a couple of passengers on the down platform and crossing over the footbridge, the tranquility is briefly disturbed as an ex Great Western Railway 0-6-0PT simmers away in the up loop platform at Princes Risborough before heading off to High Wycombe in 1955. (Adrian Vaughan)

Ex Great Central N5 class 0-6-2T engine No 69273 is seen awaiting the right away signal from Monks Risborough and Whiteleaf Halt on the 1st July 1953. (Authors Collection)

Between trains at the very charming Little Kimble station on the Princes Risborough to Aylesbury Branch sometime in the afternoon of the 8th July 1939. (H.C.Casserley)

South Aylesbury which lay on the outskirts of Aylesbury was opened on 13th February 1933 to serve a new housing development, though sadly with the advent of the motor car and an intensive bus service in the town the little halt passed into history on the 5th Jun 1967. (Lens of Sutton)

The remains of Bledlow Bridge Halt wose shelter became a platelayers hut when the Watlington line lost all its passenger services on 1st July 1957. (Lens of Sutton)

Wainhill Crossing Halt was one of four similar halts on the Watlington Line from Princes Risborough which were capable of accommodating a one carriage train and each boasted a little hut like the one shown in the shown in the picture. (H.C.Casserley)

Bledlow station was the first major port of call on the Thame Branch after leaving Princes Risborough and it not only boasted a reasonable service of passenger trains but also a healthy goods traffic. (Lens of Sutton)

On a rather misty day in the early part of Summer 1959, a 61xx Class No 6129 sits with its two coaches under the trainshed hoping to pick up some custom before leaving for Princes Risborough. (Courtesy M.Bromley)

Great Western
and
Great Central Joint Rly.

On Monday, April 1st, 1929,

A NEW HALT

will be opened at

ILMER

situate between

PRINCES RISBOROUGH and HADDENHAM

The undermentioned Trains and Rail Motor Cars will call at the times shewn :—

WEEK DAYS.

	6.10	a.m.	Marylebone (*7.32 a.m. Princes Risborough*) to Wotton	At 7.37 a.m.
	7M50	a.m.	Princes Risborough to Banbury	„ 7.55 a.m.
	9M48	a.m.	High Wycombe (*10.5 a.m. Princes Risborough*) to Banbury ..	„ 10.10 a.m.
	2M58	p.m.	Princes Risborough to Banbury	„ 3. 3 p.m.
	4M5	p.m.	Princes Risborough to Bicester	„ 4.10 p.m.
A	8.15	a.m.	Wotton to Marylebone	„ 8.26 a.m.
B	7M35	a.m.	Banbury to Princes Risborough	„ 8.53 a.m.
	10M15	a.m.	Banbury to Princes Risborough	„ 11.46 a.m.
	1M55	p.m.	Banbury to Princes Risborough	„ 3. 3 p.m.
C	4M20	p.m.	Banbury to Princes Risborough	„ 5C40 p.m.

SUNDAYS.

	8.35	a.m.	Marylebone (*10.2 a.m. Princes Risborough*) to Calvert	„ 10. 7 a.m.
D	8.22	a.m.	Calvert to Marylebone	„ 8.49 a.m.

A—Will leave Wotton at 8.11 and Haddenham at 8.21 a.m.
B—Will leave Brill and Ludgershall at 8.34 and Haddenham at 8.48 a.m.
C—Calls at Ilmer at 5.40 p.m. during period of daylight only.
D—Will leave Calvert at 8.19, Wotton 8.33 and Haddenham at 8.44 a.m.
M—Rail Motor Car, one class only.

The Third Class Single Fares from the new Halt will be as under :—

	s.	d.		s.	d.
Akeman Street	1	4	**Brill and Ludgershall**	1	4
Ardley	2	6	**Calvert**	1	11
Aynho Park Platform	3	1	**Haddenham**	0	4½
Banbury	3	10	**King's Sutton**	3	5
Bicester	2	0	**Princes Risborough**	0	4½
Blackthorn	1	9	**Wotton**	1	1

Parcels (except Milk in Churns) and Goods Traffic will not be dealt with at the Halt.

March, 1929.

By Order.

A handbill issued by the Great Western and Great Central Joint Railway proudly announcing the inauguration of a new halt at Ilmer. By the looks of it there were five services each way on weekdays and just one on a Sunday. (Courtesy D.Edwards)

Haddenham Station under construction in 1903 showing the platforms at a rather advanced stage with some evidence of work in progress as the tiny contractors engine heads along with some wagons to pick up some more materials. (Lens of Sutton)

On the 20th July 1946 a Princes Risborough to Banbury auto train is seen negotiating the rather fascinating Ashendon Junction. (H.C.Casserley)

Rather secluded in the wilds lay Wotton which was already served by a station. Here the platforms seen in the final stages of construction were of basic design as indeed was the signal cabin but down at road level a more substantial building was erected to house the Booking Hall etc. (Leicester Museums)

An excellent view of Akeman Street as seen from the Wotton side of what is now the A41 road. Note the two levels for the station whose Booking Hall and Ticket Office were in the more substantial building below right while top left is the platform and a shelter. (Lens of Sutton)

Auto trains were the forerunner of the modern bus services and here a Great Western Railway 0-6-0PT No 5419 is caught on film after having trundled through the tunnel at Brill and Ludgershall on the 22nd June 1935. (H C Casserley)

On a warm summer day in 1935 a Birmingham train thunders along through the beautifully wooded cutting towards the short tunnel that lay just south of Brill and Ludgershall station. Note the attractive clerestory coach behind the locomotive. (H C Casserley)

Ex Great Western Railway 4300 Class 2-6-0 No 9316 rushes through Brill and Ludgershall station with a mixed goods containing a fascinating variety of wagons bound for Old Oak Common. (D.Edwards)

Blackthorn Station which lay on the former Great Western Birmingham Main line just south of Bicester. Sadly due to its isolated location few passengers used the attractive facility and it was abandoned after only a short life in 1953. (Lens of Sutton)

Bicester North station from the Banbury end showing the Signal Cabin, up side platform buildings, track layout, footbridge and water column beside which is seen Great Western Railway 4900 class 4-6-0 No 5933 Kingsw ay Hall at the front of a Birmingham train. (H.C.Casserley)

Banbury General station was a neat and compact affair as seen in this January 1956 picture, perfectly adequate for the traffic it handled, though a year later it underwent a complete transformation to become one of the most modern facilities of it's time. (Courtesy M.Bromley)

A British Railways leaflet advertising engineering work on the section between Ardley and Aynho Park. (Author)

ALL CHANGE IN THE CHILTERNS

Until the early 1930's apart from the grouping of the smaller railway companies into four large organisations which saw the Great Central become part of the London North Eastern Railway, the Great Western being made the basis for a much larger group of the same name and the London North Western being swallowed up by the London Midland Scottish Railway in 1923, essentially trains were still the principal means of moving both goods and passenger traffic in Buckinghamshire and Oxfordshire. However behind the scenes the increase in the number of motor cars and lorries was gradually making its presence felt as a couple of stations fell victim namely those at Akeman Street on the 7th July 1930 and South Harefield on the 30th September 1931, two stations which generated insufficient patronage. Whilst their loss was insignificant, some of the small lines like the Brill Branch were beginning to carry fewer or no passengers, while goods traffic also was drying up. Indeed such a situation was serious enough but would not have been wholly disastrous if the London Passenger Transport Board had not been created on the 1st July 1933 which took over responsibility for all the Underground lines and Road Transport in the capital.

Immediately after this organisation was established, a programme of new works was instigated at massive expense both to modernise and convert the Underground system to a purely passenger carrying railway. Consequently in the case of the Metropolitan, the goods engines were sold to the London North Eastern Railway to continue the service, while plans were afoot to electrify the line as far as Amersham only, since the London Passenger Transport Board did not believe that anywhere beyond could justifiably be advertised as the Underground, nor indeed merited modernisation.

In this light any line with insufficient traffic or revenue north of Amersham would be bound to be considered a good candidate for closure. Sadly all too soon, the London Passenger Transport Board sent a rather brief letter on the 1st June 1935 to the Oxford and Aylesbury Tramroad Company who owned the Brill Branch to say that they would not renew their operating lease. Once the die had been cast much effort was made to hasten it's eventual end. Barely two months later on the 23rd July 1935, some directors and engineers from the London Passenger Transport Board ran a special train to assess the state of the Brill Branch which they deemed unsuitable for retention without major investment which was not justifiable considering the possible revenue. After this the closure set for the 1st December 1935 coinciding with the Metropolitan and Great Central joint working agreement expiring. In the event the final run along the branch was on the previous day with a packed train which was watched by large crowds of spectators at each station who turned out despite the appalling weather. As the oil lights went out at Brill amidst a chorus of Auld Lang Syne from those present, staff prepared the train, whilst others hastily loaded it up with every piece of removable station furniture as well as the contents of the offices, a process which was repeated at all of the other former ports of call on the journey to Quainton Road. Anything that remained was then subsequently auctioned by Messrs Burrows and Bradfield of Thame who recouped the sum of £72 7s 0d while the Ward Scrap Metal Company of London paid £7000 for removing the rails prior to them being melted down.

No sooner had the Brill Branch passed into history, than closure notices began to appear for the Quainton Road to Verney Junction line whose sad demise was scheduled to occur on the 6th July 1936 the same day that Waddesdon (Manor) would cease to be part of the Metropolitan network. Whilst passenger services actually did end on that particular day, the rusting rails remained in situ until the 28th

January 1940 when the down line was lifted from Quainton to Winslow Road. From Winslow Road up to Verney Junction the line became a long siding. The up line however was to enjoy a further seven and half years service carrying goods for the first two and then in 1943, probably due to the war, some passenger trains were run beyond Aylesbury to Verney Junction again. Once demand on the line from the war ceased, the remaining regular traffic began to dwindle away, such that goods trains ceased on the 6th September 1947 with passenger workings following shortly afterwards on the 31st May 1948, six months Britain's Railways were nationalised.

Apart from the four independent companies being made part of the new British Railways, very little happened in the next eight or so years apart from the closing of Ruislip Gardens, Blackthorn and the underused Wotton station on the 21st November 1948, 8th June and 7th December 1953 respectively. The picture was not all gloom and doom though, as in 1956 in the wake of an ever growing popularity of the motor car coupled with the increased number of bus/coach services in not only West Hertfordshire but also South Buckinghamshire, the London Transport Executive as it had become known, reeling from still having to soldier on valiantly with a system worn out and damaged by the Second World War, resolved to revive part of their 1935–1940 New Works Programme under which the surface Metropolitan Lines would be completely rejuvenated. Within the above scheme, the primary objective was to widen the trackbed from Harrow North to Watford South Junction in order to accommodate four instead of two tracks and to extend four rail electrification over the ten mile section from Rickmansworth to Amersham as well as Chesham, both of which would offer several advantageous spin offs like those listed below.

1. Increased Running Capacity Allowing Regular Interval Services.
2. Abolition of the Electric–Steam Changeover at Rickmansworth.
3. Faster Running Times and Removal of Unnecessary Delays.
4. A Smoother Interaction Between the Metropolitan and British Railways Eastern Region trains.
6. Potential For Expanding the Services Still Further.

Although the majority of the initial civil engineering works for the new cuttings, embankments and bridgeworks had been instigated prior to the outbreak of the Second World War, several structures still awaited attention by the time construction commenced in the early part of 1958. Some fifteen bridges comprising ten rail and five road bridges required widening or reconstruction. Also three stations would be affected by laying the two new tracks between Harrow North and Watford South Junction. Pinner was the first of these having to be rebuilt to allow sufficient space for the new through lines to pass by on the western side of the station, then came Northwood whose whole facility was swept away except for the old southbound platform which was switched round and redeveloped to serve Northbound Slow Trains directly opposite which a mirror facility was built for the equivalent Southbound workings, which just left Moor Park. Here the old station lay on the western side of the formation, so the London Transport Executive decided that to get the best use out of the space, two island platforms would have to be constructed. Work was concentrated on establishing the one whose platform faces would serve the Watford Branch to begin with. On completion this would be used for Rickmansworth trains also, thus allowing the demolition of the old wooden up and down main platforms and the severing of the tracks through them. Once the site was cleared, the second new island platform at Moor Park took shape quickly followed by the realignment of the main lines, after which time all attention was focused on the Watford South Junction to Amersham and Chesham section.

Without doubt the process was considerably speeded up by the fact that the London Transport Executive decided to modify the scheme slightly by altering the original plans, such that they would no longer be wanting four tracks to run from Watford South Junction to Rickmansworth, nor did they feel it was necessary to rebuild the latter or Chorleywood Station which would have resulted from the installation of passing loops.

Effectively the second part of the modernisation programme began from Rickmansworth and from here during the Summer of 1958, some eight thousand concrete posts were driven in to the ground at the side of the line to carry high tension cables supplying traction current at 600-Volts d.c. from the proposed electric substations at Chorleywood and Chalfont. Additionally these posts would carry cables linking the latest colour light signalling with two modern push button operated signal cabins at Rickmansworth and Amersham.

Having safely housed the miles of cabling required, the engineers' next task was to turn their attention to modifying the trackwork. At Rickmansworth the locomotive changeover spurs were lifted and a number of new electrified sidings were provided that were more than capable of storing the new eight car trains. Chesham however was to become more important as a terminus as it was afforded the honour of having a four coach bay road installed at the southern end of an extended main platform. Extensions to the platforms on a larger scale also occurred at Chalfont and Latimer, Chorleywood and Amersham which was soon to become the most northerly outpost of the Metropolitan. Due to the latter fact, of all the stations on the Modernisation Section, Amersham was to change by far and away the most significantly. Here an additional face was to be created on the northbound platform by doubling its width, while its length also was to increase to five hundred and twenty five feet, specifically designed for handling Aylesbury trains. Thus the lay-by siding at the south end had to be repositioned. Where this new facility diverged, the engineers installed a reversing crossover similar to that at the north end of the station, whose main difference was that two four hundred and fifty foot sidings were connected to it. In the course of the latter undertaking two road bridges had to be rebuilt as well as widened.

Once all the trackwork had been established, work was soon able to begin in earnest on laying almost thirty miles of conductor rails on both the main running lines. To begin with these rails started life as sixty foot lengths, though prior to arrival on site, they were taken to Willesden Green to be welded together into three hundred foot sections. After this they were loaded on a special five wagon train and taken to their final resting place at which point they were unloaded, positioned and welded in half mile lengths, having been bolted firmly to the sleepers. Eventually after four years of dedicated work which had involved an estimated outlay of £3,500,000, the project came to fruition on August 15th 1960, an event celebrated by the running of a crew training special of Metropolitan T stock into Chesham.

Officially the brand new regular interval, even time timetable heralding the arrival of the electrics was due to start for the general public on 12th September 1960, but because of the delay in placing the order for rolling stock, Cravens Ltd had not been able to complete sufficient of the new, aluminium, eight coach, A60 sets, leaving steam haulage to continue working some of the services from Rickmansworth to Aylesbury. This practice did not continue much longer however, as a packed farewell special hauled by Fairburn 2–6–4T No 42070 signalled the demise of Metropolitan Steam, just under a year later on the 9th September 1961. On the same day the four stations after Amersham namely Great Missenden, Wendover, Stoke Mandeville and Aylesbury, became the property of the Western Region of British Railways being serviced by regular trains from Marylebone.

Marylebone and the lines radiating out from it had also not gone without change whilst the Metropolitan electrification programme was in progress, but sadly these were to prove to be detrimental to the old Great Central and Great Western Railway network. The first difference occurred on the 1st July 1957 as the Watlington Branch lost its passenger services after which time goods trains continued to run though these were gradually cut back to Chinnor to serve the cement works. A year later in 1958 the network had to endure a strange transfer from the Eastern Region to both the Midland and Western Regions. These two organisations took on the responsibility for the sections from Ashendon and Quainton Road as far as Chesterfield and Marylebone to Banbury and Aylesbury via Princes Risborough (later incorporating the short Amersham to Aylesbury section). Within no time at all, the Western Region resolved to remove West Wycombe station from the timetable on the 3rd November 1958 in the face of increased competition from coach services. However its passing was to be overshadowed by the new regional control which brought considerable administration problems due to the frequent contact that had to be made between Paddington and York, so wasting large amounts of time. Operating practices also became difficult as many former Great Central men became confused about who they actually worked for and found it hard to adjust to any new motive power that appeared at Neasden.

No sooner had life settled down after the upheaval caused by the change in regional control, when another blow was delivered with the rather ominous axing of both the popular "Master Cutler" and "South Yorkshireman," trains, which often afforded an impressive sight whilst they stormed through Harrow, Amersham and Aylesbury or Wycombe, Princes Risborough and Ashendon on their way north. Such a premature demise of what were the line's flagship trains, only encouraged people to speculate that a deliberate policy was being enacted which would eventually see services going downhill all the more. The latter prophecy could not have been closer to the truth as within two years, on the 2nd January 1960, British Railways decided to downgrade its Great Central express services. Effectively High Wycombe lost all its through services, while in the case of Harrow on the Hill and Aylesbury, these stations now were served by three semi-fast workings in each direction instead of the long distance trains during the day. Another radical move had hit the "London Extension" which proved to be devastating as the new services were hardly attractive, offering no refreshment facilities and calling at between seven and nine stops en route, thus discouraging several regular users or occasional passengers.

However around this time, the local services in the area managed to receive a boost with the introduction of sixty Diesel Multiple Unit cars which later became designated Class 115 and 186. These were formed into four car suburban sets comprising of two Derby Works B.R. Non Gangwayed, Motor Brake seconds powered by two BUT (Leyland Albion) 6-cyl horizontal engines producing 230 bhp, with two trailer vehicles sandwiched in between, one of which contained toilets and a first class compartment. To celebrate their arrival a number of the stations received a few cosmetic improvements and had their platforms lengthened to be able to take eight cars.

Without doubt the next major development on the Chiltern Network was to bring about some hope with the pathing of the spectacular Blue Pullman Diesel Multiple Units through "Wycombe". These new trains decked out in a blue livery with a white window band, ran from Paddington to Wolverhampton Low Level, but passengers would frequently complain of their poor ride quality.

Timetable of Birmingham Pullman
(introduced 12th September 1960)

Wolverhampton (Low Level)	dep	07.00	—
Birmingham (Snow Hill)	dep	07.30	14.30
Solihull	dep	07.40	—
Leamington Spa (General)	dep	08.00	14.55
London Paddington	arr	09.35	16.30
London Paddington	dep	16.50	12.10
Leamington Spa (General)	dep	18.19	13.34
Solihull	dep	18.44	—
Birmingham (Snow Hill)	dep	18.55	14.00
Wolverhampton (Low Level)	arr	19.20	—

From 11.9.61

Wolverhampton (Low Level)	dep	—
Birmingham (Snow Hill)	dep	13.00
Solihull	dep	—
Leamington Spa (General)	dep	13.25
London Paddington	arr	14.55
London Paddington	dep	10.10
Leamington Spa (General)	dep	11.34
Solihull	dep	—
Birmingham (Snow Hill)	dep	12.05
Wolverhampton (Low Level)	arr	—

In addition to the Blue Pullman sets, the Western Region began to introduce a whole fleet of cross country units for an outer suburban service from Paddington. Meanwhile Marylebone's small concourse reverberated with the roar of Diesel Multiple Units which drowned out the once familiar sounds of the much revered steam engine in charge of a long rake of motley coaches. South of Aylesbury and Banbury by June 1962, the month that Neasden shed closed and was replaced by a new Diesel Depot at Marylebone formed out of the old Great Central Carriage Shed and sidings, steam was almost banished except for work on the semi fasts and the local services to Woodford Halse, Rugby, Leicester or even Nottingham, known as "Ords". Even these were not safe from the incursions by the Diesels which occasionally ventured north of Aylesbury on service trains, two regular workings being 4.30pm ex Aylesbury which returned at 5.10pm from Quainton Road and the other being the 5.22pm sx from Marylebone which returned from Quainton at 7.25pm to Neasden Sidings.

Soon further services in the Chiltern area came under scrutiny in terms of profitability and potential in a report prepared by Dr. Beeching who had been engaged by British Railways to reshape an ailing system to the extent that the end result would bring forth a streamlined, rationalised, cost efficient concern. From the information compiled, numerous alterations would have to be made to the services running out of Marylebone, but these would mainly be confined to the sections of the railway beyond Princes Risborough and that north of Aylesbury. The latter involved the Banbury and Oxford (via Thame) routes, which under the proposed recommendations, would see the end of all the wayside stations and a closure to all passenger traffic respectively, while the route north from Aylesbury would see the cessation of all local trains to Rugby, leaving just the Semi-Fasts. Although Beeching had seemingly been the originator of such proposals, as far back as November 1960, both the London Midland Region and the Western Region of British Railways, did actually carry out some censuses to investigate the popularity of their respective lines in the area. The Midland Region used Quainton Road as an example where they discovered that on average some 67 people caught the train while 40 left it here, the most popular departure being that at 8.26a.m. which obviously attracted school children and commuters to patronise the line. On the Banbury section, passenger traffic was very light indeed, while the Thame to Oxford route seemingly offered some potential. However in the figures produced by the Western Region they managed to show that the average number of passengers joining at Princes Risborough on a weekday was almost 40 while those alighting was 43. On Sundays these figures were substantially less. All in all the line was losing some £26,000 per year.

By mid 1962 the British Transport Commission felt it had adequate grounds for implementing its proposals as despite looking at ways of paring down costs by reducing stations to unstaffed status and introducing Diesel traction, all three local services were really so unprofitable that the large gap would never be satisfactorily bridged between income and costs. With this in mind, the concept of closures, was handed over to the East Midlands Transport Users Consultative Committee, with the relevant findings. In hardly any time at all the latter organisation came to the conclusion that a series of replacement bus services would be a sensible option and that closure could go ahead if British Railways endorsed the findings. Would British Railways do anything else! Thus the die was cast, allowing the first phase of the operation to come into effect on 7th January 1963 when Ilmer Halt, Haddenham, Brill and Ludgershall, Ardley, Aynho Park, Bledlow, Towersey Halt, Thame, Tiddington, Wheatley, Horspath Halt, Morris Cowley and Littlemore passed into History.

The services north of Aylesbury were a little more fortunate in that they continued for a further two months until the 4th March 1963. After this time Quainton Road, Calvert, Finmere and Helmdon all ceased to play host to passengers but trains still continued their wearisome journeys north of Aylesbury to Brackley, Woodford Halse, Rugby, Lutterworth, Ashby Magna, Leicester, Loughborough, East Leake and Nottingham. Ultimately even these were to be axed, though the slimming down of railway activity in the Chiltern area underwent a brief respite period which was only disturbed by the closure of Calvert goods yard on 4th May 1964.

By the Autumn of 1964 this lull reached an inevitable conclusion with the all too sudden announcement that from the 7th September 1964 the Banbury to Woodford Halse railmotor services were being withdrawn. Shortly afterwards, lack of custom was to bring about the end of two more goods depots at Finmere on 5th October 1964 and Helmdon which bowed out on the 2nd November 1964. Without a doubt, the once grand Great Central London Extension was now on a downward course, leading to its eventual closure. Essentially there were several reasons for this, the primary one being that the line was unprofitable and showed little scope for improvement in receipts as it ran through areas of scant population. Secondly it was labelled as a duplicate route, because after 1958 when it became part of the Midland region of British Railways, the latter body already served Rugby, Lutterworth, Leicester, Loughborough and Nottingham with their other main lines, leaving only those places of no consequence like East Leake, Ashby Magna, Woodford Halse and Brackley with the distinction of being unique ports of call for the remnants of the Great Central. Thirdly traffic was able to be diverted away easily onto other lines.

And so during the early Summer of 1965 on June 12th and 13th, all the goods and several of the cross country passenger trains were rerouted to run via Oxford and Birmingham. A number of these used to head over to Oxford from the joint line at Princes Risborough, via Thame but alas the practice ceased when the section between the latter and Horspath was officially closed later in the year. Traffic now became less and less on the Marylebone network such that mounting pressure was now being brought to bear on removing the remaining Nottingham Semi-Fast workings, to allow closure of the Aylesbury to Rugby section and the creation of a much smaller streamlined system that would suit the requirements of Suburban commuters and off peak travellers. Soon despite a modest protest about the hardship that loss of the services would cause, during which time Quainton Road ceased to handle any goods traffic, Mrs Barbara Castle, the Transport Minister gave her authorization for complete closure of the London Extension. Although the execution warrant had been signed in early July, the final trains did not run until the 3rd/4th September 1966. Over this weekend, unlike the death of a railway today, few people were there to mourn its passing. Trainwise apart from strengthening the regular 17.15pm Nottingham to Marylebone and the 22.45pm Marylebone to Manchester workings with extra coaches along with a Locomotive Club of Great Britain special from Waterloo to Sheffield and back hauled by a mixture of former SR Merchant Navy No 35030 Elder Dempster Lines and two ex LNER B1's, the Great Central was to fade away quietly.

Meanwhile after the passing of the Great Central, once the track lifting was progressing at a rapid rate, plans were being drawn up to reduce operations on the line through Wycombe to the West Midlands and beyond. For many years this section had always had the proud accolade of being the shortest line between London and Birmingham with a journey time of two hours, however it had one fundamental flaw in that it passed through fewer places of great significance than its longer rival route through Slough, Reading, Didcot and Oxford. Additionally once electrification of the West Coast Main Line was complete, most of the trains to Merseyside, the West Midlands and the North Wales Coast would be switched to run out of Euston commencing on the 6th March 1967. Thus it was that the Chiltern Hills reverberated for the very last time with the sweet sounds of those despairing old steam workhorses, while they hauled their rakes of carriages on trains like the Cambrian Coast Express and the Inter City. However the ceremonial farewell was not confined to these prestigious trains alone as two special workings were put on from London Paddington to Birkenhead, headed by two beautifully polished locomotives in the shape of ex Great Western Railway 4-6-0 No 4079 Pendennis Castle and sister engine No 7029 Clun Castle. Exactly three months after, around the time South Aylesbury Halt closed on 5th June 1967, tracklifting teams moved in very quickly to single the double track section between Princes Risborough and Aynho junction as well to remove all the track on the Ashendon Junction to Grendon Junction portion save for a short siding serving Akeman Street from the latter. Due to a change in its status, the Princes Risborough to Aynho Junction route was realigned for faster running and treated to some modern colour light signalling. After the rationalisation works had been effected on the latter, the contractors moved over to the Oxford line to lift the Thame to Horspath Halt section in 1969, a year after the Oxford to Bletchley route was downgraded to a freight only line.

Meanwhile further down the Wycombe, Railway traffic had steadily been declining since the early 1960's on the section from High Wycombe to Maidenhead such that several economies were made to reduce costs which included the introduction of Diesel Multiple Units on a reduced Maidenhead to Aylesbury through service and the downgrading of the stations at Wooburn Green and Loudwater which were reduced to Paytrain Halts. Sadly things did not pick up and in a rather unusual move in this era rife with closures, British Rail decided to give the line one more chance, cutting the trains back from Aylesbury to High Wycombe. However little difference occurred and on the 4th May 1970 with no time being allowed for the public to voice their opinion, the Bourne End to High Wycombe line was closed and the track quickly lifted.

In the 1970's the pattern of services on the truncated Chiltern network settled down to offering a fairly intense peak service with a locomotive hauled express departing from Birmingham New Street for Paddington at 06.40am and returning at 17.40pm. Off Peak an hourly service ran from Marylebone via Amersham as far as Aylesbury with an identical pattern from Marylebone to High Wycombe. North of High Wycombe trains ran two hourly to Banbury with the odd one diverging to Aylesbury from Princes Risborough. Apart from Passengers, a number of freight services continued to use the Wycombe line, these principally carrying minerals, oil, parcels, cement and general produce on the Chiltern Trader. By 1972 the Chiltern Trader disappeared along with the parcels trains into Marylebone while in 1977 Brick traffic ceased to be shipped from Calvert. Instead this traffic was replaced by the shipment of waste from Northolt in giant container trains which dumped their contents into the massive pits left by the extraction of materials for making bricks. Such guaranteed healthy traffic was very welcome indeed and meanwhile increased flows occurred with more oil to Thame, more cement from Chinnor, extra wagons of fertilizer to Akeman Street and greater parcels, coal and oil movements at Aylesbury. Meanwhile the picture was not so rosy at High Wycombe goods depot which closed completely to all traffic.

At the beginning of the 1980's nothing really exciting seemed to be happening on the lines out of Marylebone, that was until a new report outlining a five year strategy for British Rail, persuaded managers that certain cuts would have to be made for the complete system to become profitable by 1988. On the London Midland Region the bosses would definitely be focussing their attention towards Marylebone which had been under scrutiny before. The reasons were simple in that the track on this tiny network was in a poor state of repair, the lineside equipment and signalling was ludicrously antiquated, while the rolling stock had been operating for almost thirty years. In essence the system needed a total refurbishment but the money was not then available, so the only logical option seemed to be to make cutbacks which caused a minimum of hardship but still maintained a basic service.

The concept entailed the diversion of all Banbury, High Wycombe and the Aylesbury (Peak Service) trains into Paddington on the one hand, whilst on the other, the management proposed handing over the Amersham to Aylesbury section to London Underground's Metropolitan Line in order that they could single the route and electrify it using the fourth rail system. Additionally British Rail tracks would disappear between Northolt Park and Harrow on the Hill as far as Marylebone. Marylebone itself offered a real incentive as if the station could be closed, then once British Rail moved its administrative functions from the former Great Central Hotel to Euston House, the complete site could be sold off for an estimated £10 million.

British Rail subsequently maintained that all this was purely a rumour and that they had no intention of closing Marylebone, but history told a different story as two previous attempts had been made to divert traffic into Paddington and hand over the Amersham services to the Metropolitan. Naturally it wasn't long before the newspapers got wind of the story and in the Bucks Free Press, a huge article appeared entitled "Marylebone Again Faces The Axe – The End Of The Line For The City Commuter." Once the story broke it became of wide spread interest and fears were growing that the battle of Marylebone was just beginning. At this juncture, a well known pro-rail organisation, the Railway Development Society got deeply involved. Within no time at all, the latter became a major force in establishing a campaign to prevent the ultimate closure of Marylebone and its associated lines. An inaugural meeting was held in the first week of October 1983 at Aylesbury Civic Centre, where a packed crowd declared its unanimous objection to British Rail's proposals, demanding instead a programme of improvements.

With the regular commuters and occasional travellers stirred up, they actively began investigating ways to protect their railway. To this end, a meeting was held at the Old School Hall in Great Missenden where a motion saw the birth of the new Aylesbury and District Rail Passengers Association whose aims were to fight the closure of Marylebone, prevent any deterioration in services and in order to present a united front to consider a link up with a few of the Metropolitan Line Pressure Groups who were faced with problems relating to a proposal to close the Chesham Branch not only due to the limited number of passengers using it, but also because an inordinate amount of expenditure would be required to replace the two bridges at Waterside and Moor Road which neither London Transport nor the council was prepared to fund.

Whilst the saga of who was going to finance the bridges raged, the Aylesbury and District Passengers Association began to make plans for running a "Shoppers Special" train to Milton Keynes on the 9th June 1984 to prove how popular an extended service from Marylebone could be and to show that a justifiable case could be presented for reopening the freight only line north of Aylesbury. At around the same time due to a major mishap in the confines of Paddington station, Marylebone suddenly found itself useful when for four days it had to play host to the Paddington to Birmingham New Street expresses.

In the meantime the controversial document known as the "Serpell Report" was published advocating the conversion of unprofitable railway lines into roads, which prompted the National Bus Company to engage Professor Peter Hall (a leading transport expert from Reading University) to carry out a feasibility study for creating a dedicated coachway leading out of Marylebone through Lords and West Hampstead Tunnels until it could form junctions with the A40 and M25 roads. The National Bus company believed that should such a scheme be successful then up to an hour would be taken off express coach journey times into the capital and some 250,000 extra trips would be taken off the roads of London overnight. Although the proposal gained a fair amount of momentum, in a matter of months Chiltern commuters were confronted with the appearance of a further report which Coopers Lybrand had been commissioned to prepare during the previous year. However the latter company waited to reveal their findings until the appropriate moment which turned out to be some twelve months later. On the one hand there was good news in that the report effectively put the dampeners on the Express Coachway scheme but on the other hand instead of retaining Marylebone as a station, they recommended the sale of the site, thus allowing British Rail an opportunity to recover a possible £50 million for investment in another part of the network. This was all very well but what would travellers do to get to work without a convenient station?

After the report's publication, a tremendous furore developed to the extent that the Marylebone Travellers Association, which had lain dormant for many years convened an emergency meeting at the Beech Tree public house in Beaconsfield which was attended by just over a hundred people. During this meeting an important benefit evolved as the editor of the Buckinghamshire Advertiser launched an official "Save Our Station Campaign" which would be a weekly feature until Marylebone and all the other threatened parts of the Chiltern Line had a secure future.

By the summer of 1984 the campaign had stepped up with many more parties becoming involved who not only helped circulate a barrage of anti-closure literature but also actively made the travelling public aware of a worsening situation. Much to the delight of the Aylesbury and District Rail Passengers Association, their special train to Milton Keynes ran full on the 9th June 1984, proving not only was the retention of the Marylebone lines important but also that scope existed for reinstating passenger services on the line north of Aylesbury to Milton Keynes.

Throughout July, August and September matters reached a stalemate situation that was until a meeting to excel all others occurred at Beaconsfield Town Hall on the evening of the 24th October. The platform speakers on this particular occasion were Sir Ian Gilmour MP for Chesham and Amersham and Mr.Gordon Pettitt British Rail's Deputy Director of London and South East Services. The principal aim of the meeting which was attended by a large crowd of angry protestors was to outline to Mr. Pettitt what their feelings were in the hope that he would convey such a huge weight of objections back to Rail House. For its second strategy the Marylebone Travellers Association wanted to organise a meeting of all the commuter groups in the campaign in order to pool their ideas and avoid conflicting views being put across in public whilst arguing the case with British Rail. During the long meeting the assembled audience was also informed that a spokesman for the London Regional Passenger's Committee who were ultimately responsible for looking at the hardship closure would cause, said that some six hundred letters had been lodged by the 8th October which was the deadline for objections. However no date had been chosen for any inquiry or public hearings.

Once the Christmas period had passed and the New Year was in full swing around the time the "Concrete Cow" railtour ran on the 12th January 1985 which not only had a peculiar formation with a Class 33/1 at either end and 4TC 421/401 but also worked an incredible circular schedule from Watford Junction which took in Kensington Olympia, Clapham Junction, Barnes, Kew East Junction, The Neasden North and South Junctions, West Ruislip, High Wycombe, Aylesbury, Quainton Road, Bletchley, Milton Keynes Central, Bicester London Road, Oxford, Reading, Ash, Guildford, Redhill, East Croydon to name but a few places, an announcement was made that the inquiry into the closure of Marylebone and both the Neasden to Northolt and Neasden to Harrow on the Hill lines would be in three parts, the first being at Rickmansworth on the 25th February while the venue for the second and third sessions would be the magnificent Great Western Royal Hotel at Paddington. Now for the protestors it seemed that there was a light at the end of the tunnel, that was until the closure hearings were deferred until June.

Suddenly all attention turned away from the closure campaign with rumours becoming rife that Marylebone would play host to a steam working organised by the Steam Locomotive Operators Association, who had long been pressing British Rail to allow steam back into London. Eventually the wait was over as on the 12th January 1985, a large crowd had bravely turned out despite the cold conditions in order to witness a strange piece of history. Soon everybody's excitement began to increase and the suspense heightened with a regular series of loud speaker announcements announcing that the train was at High Wycombe, Beaconsfield, Gerrards Cross, Denham, Wembley and Neasden after which with a loud blow on its whistle A4 4-6-2 No 4498 Sir Nigel Gresley in charge of The Thames Avon Express emerged out of Lords Tunnel to be greeted with celebrity status as it sidled into Platform 4. The latter engine's arrival was very convenient in that it would allow British Rail a chance to satisfy the needs of one of its largest customers namely the Post Office who were wanting to launch their Famous Train Stamp Issue. The stamps themselves depicted five scenes by the famous Terence Cuneo these being the "Golden Arrow", "Cheltenham Flyer", "Cornish Riviera", "Royal Scot", and "Flying Scotsman". Two big panels with these stamps on were fixed to "A4" 4-6-2 No 4498 Sir Nigel Gresley which trundled around Marylebone in light steam on the 21st January, while the second was attached to "47" No 47500 which headed a VSOE Pullman special from Paddington over to High Wycombe and Banbury before returning via Oxford and Didcot.

After the success experienced with the Post Office train, British Rail resolved to encourage the running of further steam specials out of Marylebone in association with the Steam Locomotive Operators' Association who used "A4" 4-6-2 No 4498 Sir Nigel Gresley on the first few trips on the 26th January and the 3rd and 16th February respectively during which time a catering service was introduced and the service received much publicity from the BBC's Mike Reid and Sarah Greene. Naturally behind the scenes with the new steam services there were a number of teething troubles requiring minor refinements like adjustments to the services, provision of better watering facilities, easing off on timings and controlling of the biggest "Bête Noir" the trespasser. Once all these problems were seen to, a different locomotive in the shape of a Merchant Navy Class No 35028 Clan Line was brought in to work the popular steam specials until Tuesday the 30th April. On this day she performed the outward leg of an under patronised evening trip to Stratford Upon Avon which returned to London pulled by D200. The following Saturday the beautiful and powerful Princess Coronation Pacific No 46229 Duchess of Hamilton stormed down from Sheffield with a South Yorkshireman in readiness for doing a few Stratford turns.

With the appearance of the Duchess of Hamilton and the obvious success of the trains which had already run, British Rail began marketing and operating it's own "Shakespeare Limited" specials which offered a tour to Stratford Upon Avon with a superb roast beef luncheon on board. Unfortunately the Duchess appeared for just three turns of duty on the 12th, 25th and 26th May before disgracing herself by causing damage to a bridge. Shortly after on the 8th June she was towed to Aylesbury to make an appearance at the celebrated Rail Week. As a result No 35028 Clan Line was left to continue on all rosters from the 9th June until lowering of bridge 17 allowed No 46229 to resume working which turned out to be the 21st June.

Meanwhile it was becoming increasingly obvious that the success of the steam specials had had some bearing on the fate awaiting

Marylebone and its associated lines, since on the 13th May two positive steps forward occurred for both the Banbury and Aylesbury routes the first being the adoption of the catchy marketing title Chiltern Line, accompanied by a very attractive leaflet explaining that the network was geared towards offering a personal service and encouraging people to visit both London and Local tourist haunts with discount vouchers provided every time a ticket was purchased. Secondly due to the heavy pressure brought to bear on the London Regional Passengers Committee by a few London Boroughs, the GLC, Buckinghamshire County Council and the Railway Development Society, the June closure hearings set for the 17th, 18th and 21st were consequently postponed for an indefinite period. However even though these events proved a crucial about turn the battle had not yet been won.

Whilst the intricate legal wrangles continued over the station's fate, Marylebone was enjoying a welcome popularity with a steam special nearly every weekend right the way into September on the 28th of which month A4 Class 4-6-2 6009 Union of South Africa was rostered to work a South Yorkshireman excursion to London so that it could be employed on a Shakespeare Express the following day. Alas this rather hectic programme of events was not to be, since British Rail refused to grant a main line running certificate to the engine until it had had some minor attention. A similar treat lay in store for No 46229 Duchess of Hamilton which made its way home, working a northbound South Yorkshireman on the 5th October. Meanwhile Clan Line performed on two final trains prior to being taken to Southall for repairs.

Returning to the ill starred A4 Class 4-6-2 6009 Union of South Africa, a further attempt was made to bring it to Marylebone on the 1st November so that it could work a Marylebone to Stratford Upon Avon train on the 2nd November before returning northwards on the 16th. They say lightning does not strike twice though in this case it did, but fortunately the Keighley and Worth Valley Railway raced to the rescue providing a replacement in the form of West Country 4-6-2 34092 City of Wells which not only managed to feature on television, but also delighted her many followers as she put on a fine performance.In the following week an old favourite A4 4-6-2 No 4498 Sir Nigel Gresley returned to Marylebone with a Steam Locomotive Operator's Association organised South Yorkshireman heralding the beginning of a six month stay. A4 4-6-2 No 4498 Sir Nigel Gresley was given one further train to work before being put in store for the next couple of months. Soon after with Christmas looming large British Rail and SLOA had been provisionally preparing to run some Santa Specials to run out of Marylebone for which the two organisations had assembled A3 Pacific No 4472 Flying Scotsman, N15 King Arthur No 777 Sir Lamiel and Merchant Navy Pacific 35028 Clan Line. All three would be working from Marylebone to Aylesbury. Additionally moderate, but not enough interest had been shown for a Mini Santa operation running between Marylebone and West Ruislip, using the Buckinghamshire Railway Centre's 0-4-4T Metropolitan No 1 as well as their 94xx 0-6-0PT.

Early in the New Year a programme of steam excursions was worked out that was designed to run at least until Marylebone closed but in the event as the representative selection on the tables below show, the specials have kept on coming right up until recently.

Description		**Marylebone To Stratford Upon Avon (B)**
Dates	1986	5 Jan,16 Feb, 9 Mar, 15 Mar, 20 Apr,, 27 Apr 4 May, 11 May, 18 May, 26 May, 1 Jun, 8 Jun, 15 Jun, 22 Jun, 6 Jul, 13 Jul, 20 Jul, 27 Jul, 3 Aug, 10 Aug, 1 Aug, 24 Aug, 31 Aug, 14 Sep 28 Sep, 12 Oct, 12 Oct, 19 Oct, 26 Oct, 26 Oct, 2 Nov2, Nov 16 Nov, 6 Dec, 28 Dec
	1987	18 Jan, 15 Feb, 22 Mar, 12 Apr, 26 Apr, 10 May, 24 May, 31 May, 21 Jun, 28 Jun, 12 Jul, 19 Jul, 26 Jul, 2 Aug, 16 Aug,, 30 Aug, 13 Sep, 27 Sep, 11 Oct, 18 Oct, 1 Nov,
	1988	17 Apr, 29 May, 12 Jun, 17 Jul, 14 Aug, 14 Aug, 11 Sep, 2 Oct,16 Oct,
Title		Shakespeare Limited
Motive Power		(4498 Sir Nigel Gresley), (777 Sir Lamiel), (4472 Flying Scotsman), (35028 Clan Line), (4771 Green Arrow), (34092 City of Wells), (5305 Alderman E.Draper), (7029 Clun Castle), (4468 Mallard), (37 427), (47 481).

Description		**Marylebone To High Wycombe (B)**
Dates	1986	13/14/20/21 Dec. (four trains)
	1987	3 Jan, 4 Jan, 5 Dec, 6 Dec, 12 Dec, 13 Dec 19 Dec, 20 Dec, 28 Dec, 29 Dec.
Title		"Santa Specials"
Motive Power		(4498 Sir Nigel Gresley), (35028 Clan Line), (4472 Flying Scotsman), (4771 Green Arrow), (5305 Alderman E.Draper), (777 Sir Lamiel).
Description		**Marylebone To York/York To Marylebone**
Dates	1986	4 Oct (R),
	1987	21 Mar(B), 9 May(R), 18 Jul(R), 29 Aug(O),
	1988	9 Apr(B), 21 May(B), 18 Jun(B), 27 Aug(R), 29 Aug(B), 15 Oct(O).
Title		"South Yorkshireman"
Motive Power		(4498 Sir Nigel Gresley), (35028 Clan Line), (4472 Flying Scotsman), (34092 City of Wells), (5305 Alderman E.Draper), (4771 Green Arrow), (4468 Mallard)
Description		**Marylebone To Tyseley/Tyseley to Marylebone**
Dates	1986	3 Sep(E), 13 Sep(Z), 29 Nov(E)
	1987	23 May(B), 31 May(O), 11 Jul(A), 11 Sep(E), 12 Sep(J), 12 Sep(E), 21 Oct(D)
	1988	23 Apr(M), 30 Apr(C), 28 May(E), 29 May(E), 12 Jun(E), 24 Jun(E), 24 Jul(E), 13 Aug(E), 14 Aug(T), 11 Sep(E), 18 Sep, 30 Sep(E), 2 Oct(E), 8 Oct(I).
Title		A) Real Ale Special C) The Chiltonian D) BR Exhibition Train E) Empty carriaging stock I) The Inter City J) John Player Special M) Risborough Venturer T) Test Train. Z) Tyseley Connection.
Motive Power		(4498 Sir Nigel Gresley), (7029 Clun Castle), (5305 Alderman E.Draper), (35028 Clan Line), (34092 City of Wells), (4472 Flying Scotsman).
Description		**Open Day Duty**
Dates	1986	A) 22 Feb, Marylebone; B) 25 May, GWS Didcot; C) 6 Jun, Chart Leacon; D) 14 Jun, Aylesbury; E) 11 Jul, Rugby; F) 22 Aug, Cannon Street;
	1987	G) 3 May, Bounds Green; H) 30 May, Norwich Crown Point (J) 20 Jun, Aylesbury; K) 17 Oct, Ripple Lane Open Day.
	1988	L) 27 Feb, Bournemouth; M) 30 Aug, Wolverton.
Title		"Exhibition Service"
Motive Power		(4472 Flying Scotsman), (777 Sir Lamiel), (4498 Sir Nigel Gresley), (35028 Clan Line), (5305 Alderman E.Draper), (80080), (6201).
Description		**Various Test Runs**
Dates	1986	To High Wycombe (B) 11 Jan, 21 Feb, 19 Apr, 7 Sep. From Banbury 18 May, To Banbury 14 Sep. From Southall to Swindon 28 Jul, 29 Jul From Swindon to Marylebone.
	1987	To High Wycombe (B) 11 Jan, 21 Feb, 19 Apr, 7 Sep.
	1988	To High Wycombe (B) 11 Jan, 21 Feb, 19 Apr, 7 Sep
Motive Power		Various
Description		**Various One Off Special Trains**
Dates	1986	12 Jan SB, 25 Jan SB, 22 Mar AR, 12 Apr SO, 12 Apr SR, 5 May SB, 11 May SB, 16 May SO, 28 Sep WO, 12 Oct WO, 12 Oct WR, 8 Nov AO, 29 Nov SO, 30 Nov SR, 11 Dec DO, 12 Dec DR, 15 Dec M.
Titles		12 Jan Half Century Limited, 25 Jan Nicholas Nickleby, 22 Mar The South Yorkshireman,12 Apr William Shakespeare (Two Trains), 5 May and 11th May Cambridge Connection, 16 May 2060 Harrison Express, 8 Nov Peter Allen Limited.
Dates	1987	2 Apr PB, 4 Apr CO, 12 Apr SB, 26 Apr SB, 28 Apr QB, 31 May WO, 7 Jun TR, 20 Jun WR, 22 Jun MO 22 Jun YB, 24 Jun IO, 26 Jun CO, 29 Jun QR, 4 Jul SB, 19 Jul QB, 15 Aug DO, 9 Sep SB, 3 Oct QB, 3 Oct UB, 19 Oct ZB, 8 Nov QB, 21 Nov SB, 26 Nov VB, 1 Dec XO.

Titles		12 Apr Inter City, 28 Apr BP Special, 7 Jun Cromwell Tools Special, 22 Jun British Coal Special, 29 Jun The Senior Citizen's Special, 4 Jul CURC Special, 19 Jul The MNLPS Special, 3 Oct The Haddenham & Thame Pioneer, 8 Nov McAlpine Special, 21 Nov RSC Charter, 26 Nov The A4 Renaming Special.
Dates	1988	6 Feb FO, 27 Feb CR, 8 May AO, 9 May QO, 9 Jul WO, 17 Sep GO, 8 Dec CO.
Titles		6 Feb White Rose, 8 May TPO ecs, 9 May The Postal Pullman.
Locomotives		(4472 Flying Scotsman), (35028 Clan Line), (777) Sir Lamiel, (5305 Alderman E.Draper), (48151), (80080),(4498 Sir Nigel Gresley), (6201 Princess Elizabeth), (4468 Mallard), (34092 City of Wells),(75069).

B = Both Ways, O = Outward, R = Return, C = Carnforth, D = Derby, P = Princes Risborough, Q = Banbury, I = Ilford, M = Maintenance, S = Stratford Upon Avon, T = Burton On Trent, Y = West Ruislip, Z = Worcester, U = Butterley, W = Salisbury, V = Gerrards Cross, X = Hull, A = York, F = Sheffield, G = Dorridge.

Description		**Derby Midland to Marylebone (R)**
Dates	1989	27 Mar (R), 28 Aug (R)
Title		"Peaks Express"
Motive Power		(4498 Sir Nigel Gresley)
Description		**Chesham to Watford (C)**
Dates	1989	1/2 Jul, 8/9 Jul,
Title		"Chesham Centenary Special"
Motive Power		(L44 Met NO 1) (0-6-0PT 9466) (Met NO 12 SARAH SIDDONS)
Description		**Nottingham to Marylebone (R)**
Dates	1989	11 Nov (R)
Title		"Sherwood Forester"
Motive Power		(34027 Taw Valley) (Diesel)
Description		**Marylebone To Stratford Upon Avon(R)**
Dates	1989	2 Dec (R), 30 Dec (R)
	1990	31 Mar (R),
Title		"William Shakespeare"
Motive Power		(34027 Taw Valley)
Description		**Marylebone To Nottingham (R)**
Dates	1990	20 Jan (R), 21 April (R),
Title		"Robin Hood"
Motive Power		(34027 Taw Valley) (71000 Duke of Gloucester)
Description		**Marylebone To Sheffield (O)**
Dates	1990	17 Feb (O), 28 April (R).
Title		"White Rose"
Motive Power		(3442 Great Marquess) (35028 Clan Line) (44932)
Destination		**Amersham to Harrow on the Hill (C)**
Dates	1990	21/22 Jul, 28/29 Jul,
Title		"Amersham Specials"
Motive Power		(L44 Met NO 1) (Met NO 12 SARAH SIDDONS) (0-6-0PT 9466) (BR 2-6-4T NO 80080)
Destination		**Marylebone To Stratford Upon Avon**
Dates	1990	22 Dec.
Dates	1991	9/23 Feb. (Cancelled due to Engineering).
Title		William Shakespeare
Motive Power		4472; 35028.

23 FEB	LAST STEAM OUT OF MARYLEBONE DUE TO NEW TRACK AND SIGNALS IN THE MARYLEBONE AREA.
1 NOV	CONFIRMATION THAT STEAM CAN RETURN TO LONDON ONLY IF THE TRAINS WILL RUN OUT OF PADDINGTON INSTEAD OF MARYLEBONE. HOWEVER THEY WILL CONTINUE TO USE THE WYCOMBE LINE UNTIL SOME OTHER ARRANGEMENTS CAN BE MADE.

THE WAY AHEAD

As Spring 1986 came into full swing, the campaigners fighting to save Marylebone Station were agonizingly awaiting the setting of a date for a Public Enquiry which would allow them a final chance to protest against the closure and the associated alterations to railway services in the Chiltern area. Suddenly like a bolt from the blue, in contrast to every previous indication, the British Rail Board announced that far from a controversial execution, a complete reprieve would be issued on the 30th April. This proved to be a red letter day for a part of the national network that had often been butchered or threatened.

Four factors contributed to this almost unbelievable about turn in British Rail policy. Primarily the possibility of Marylebone closing had depended upon the ability of Baker Street's Surface Lines to swallow up all the former British Rail Amersham traffic, but due to a massive 15% increase in peak time patronage on the Underground since 1983 resulting from the success of the Travel and Capital Card schemes which encouraged more commuters to use the system for travelling to work, any additional influx of new passengers would have created serious overcrowding on existing services and no extra trains could at that time be worked into the Arrival and Departure schedule without a total recasting of the timetable.

Secondly common sense had again prevailed over the idea to path trains from Banbury, Aylesbury and High Wycombe into Paddington which would have added fifteen minutes extra to the journey into London and in addition interfered with the already tight arrival and departure schedules for Western Region local services. The two latter concepts in theory seemed a marvellous concept but in practice would have been far from practical.

Not only did railway practicalities play a part in the downfall of the plans for the Marylebone site, but a certain element of politics entered into the decision. The first of these slightly less well publicised reasons was that it appeared pointless to abandon the site when there was no immediate hope of being able to realise capital from its sale, but more importantly British Rail was just establishing Network South East whose parameters were to clean up, improve, modify and expand the facilities and services in the Home Counties, East Anglia, The Coast and Kent. Under such a manifesto removing a portion of this newly created network would therefore have been embarrassing and very untimely.

In the event the retention of Marylebone and its two associated lines to Aylesbury and Banbury, caused much jubilation amongst the MP's, Pressure Groups and Regular Travellers who felt that they had saved the day. However almost everybody forgot that a body of pressure groups were only a small thorn in the side of a large organisation who had complete control over the eventual outcome, namely British Rail. Had the latter wanted to complete their plans badly enough, then they would have starved the lines of traffic by reducing the number of trains, made the remaining ones run at awkward times and pared down the infrastructure by removing staff from the stations and rationalising the trackwork, as they had done in the days of Doctor Beeching on lines like the Waverley Route, The Great Central and Somerset and Dorset, prior to effecting closure on the grounds that they were uneconomic to continue operating.

Continuing the operation was now a problem for British Rail, as it was faced with the mammoth task of having to examine urgently ways of rejuvenating one of the most antiquated rail networks in the London area, to pull it into line with the other modernised commuter services. With scarcely a week having passed since the day when the reprieve for Marylebone was announced, British Rail acted with incredible speed to spread the good news through the popular Buckinghamshire local papers that it was now definitely committed to modernising Marylebone and improving facilities at all the stations on both routes whilst revitalising the services that called at them. However their sudden new found interest in a small network that had been left to crumble quietly over such a long period of time, naturally provoked much caution.

Nevertheless British Rail was true to its word and in order to appease the local commuters and campaign organisers, some small amounts of cosmetic work were done to the station, whilst more substantial plans were on the drawing board to catch up on the backlog of investment that was needed over the previous twenty or so years. Within a short time after much deliberation over the best way to modernise the Chiltern Line, a full report was issued outlining a £75 million programme to be completed by the Autumn of 1991. Included in this manifesto was the promise of a new set of trains, a new maintenance depot, revised track layout, a more intensive service and new signalling which would include the trials of ATP or Automatic Train Protection using the Selcab system installed by GEC Alsthom. The latter works on a track to train link with information being relayed from a signal interface cabin through an enclosure containing a telegram generator, power amplifier, selector and RF generator to the cab of the unit which picks messages up via its underfloor Antennae and sends them off to the driver.

However before the works really got into full swing a number of changes were to occur on the Chiltern Line beginning with the construction of a new station during August and September 1987 between Princes Risborough and Bicester North which would serve the catchment area of Haddenham and Thame as its name suggests. When completed this new £430,000 facility, partly financed by a £72,000 subsidy from both Buckinghamshire and Oxfordshire County Councils, boasted a huge 180 space car park, a long platform and a small ticket office with a two hourly service to Banbury and London respectively which began on the 3rd October. To mark the occasion, a special train headed by No 4498 "Sir Nigel Gresley" gently steamed into a crowded platform shortly after which the opening ceremony was duly performed. On the day after, an event of lesser proportions took place which is still worthy of brief mention, namely the transferring over of both the Marylebone to Aylesbury and Banbury lines from the Midland Region of British Rail to the Western Region. It was the latter sector which was fortunate enough to benefit again on the 5th October with the long awaited reopening of a station Birmingham Snow Hill, after nearly fifteen years absence. Hopefully, if all goes according to plan, in May 1993 it will become one of the northern termini for Marylebone trains, the other will be at Stratford Upon Avon which will be served by a a single Class 153 Railcar that will part company and rejoin its parent Class 156 Unit at Leamington Spa after running along the branch.

In taking over the Chiltern Line operations, the Western Region announced that Marylebone would be closing as a DMU fuelling and servicing point leaving Old Oak Common to take on its role, while Bletchley would be catering for major stock examinations until the new depot at Aylesbury was completed in readiness for the appearance of the successors to those faithful life-expired thirty year old Class 115 DMU's, in the form of a whole fleet of vehicles known as the Class 165/0 Networker Turbo's. Eager to be pressing on with this revolutionary design, Network South East's management submitted their designs to the Minister of Transport Michael Portillo who gave permission for an initial construction of some seventy seven vehicles at a total price of £43.5 million pounds, though it was soon found that an extra twelve cars would be needed due to an expansion in traffic.

The networkers were offered to British Rail Engineering Limited at York Works who produced eleven three Car and twenty eight two car aluminium bodied sets, specially for the Marylebone services. Unlike their predecessors they are slightly longer at 23m so as to cope with tight curves on the inner suburban lines and wider to take full advantage of the broader gauge of the Great Central. Despite having light weight body work their speed range is only five miles per hour faster than the Class 115's which they have replaced but their real advantage is in terms of acceleration as this enables the Class 165's to attain higher speeds in between stations. Consequently the journeys to and from Banbury are now taking eighteen minutes less, while those on the Aylesbury route have been pared down by 9 minutes. This is achieved due to each car being fitted out with a Perkins 2006-TWH engine capable of producing 350hp along with a slick Voith Converter and Gmeinder Final Drive.

Specificationwise, two sliding-plug saloon doors in both sides of every vehicle give the passenger access into an ultra modern interior. Once on board it is possible to see a stunning range of hitherto untried features like the driver controlled carriage dot matrix display which quite quaintly pops up with the name of the next port of call. Should the following station provide more passengers than are expected, Network South East reckons that it has banished overcrowding on the Turbo's with the advent of high density 2+2 first and 2+3 standard class seating, accommodating a total of 16 first and 170 standard in two cars and 24 and 262 in three cars respectively. Additionally a long awaited benefit has been the introduction of an interconnecting gangway between carriages. All in all the ambience that comes across is a rather inviting one for the passenger, with the roominess and comfort that is a prerequisite for this day and age.

Whilst a lot of energy has obviously been expended in the latter area, a massive amount of attention has been devoted to safety. Black box recorder fitments are incorporated in each unit, made by the Italian firm Parizzi, allowing information to be supplied on any unfortunate incidents. One problem that exists stems from the fact that the Class 165's have to run along the Metropolitan system between Harrow on the Hill and Amersham, so trip cocks are fitted to each train to prevent a SPAD (Signal Passed At Danger) situation.

Whilst all this technical wizardry was being hurriedly installed in the Networker Turbos, a rapid programme of works started to ensure that the permanent way was made ready. Fortunately due to the nature of the eastern route, which had been constructed with two tracks and no thrills as a suburban railway, it meant that only the terminal points on the Marylebone to Aylesbury via Harrow on the Hill and Amersham have required major track alterations. Due to the latter's compact nature in the early years as now, it was a route that often ran with insufficient capacity for any extra trains. Consequently as has been mentioned earlier, in the late 19th century with the advent of Great Central Railway expresses, the Metropolitan soon became overstretched such that a new means of reaching London Marylebone had to be found. The only way this could be achieved was by the Great Central effecting an alliance with their neighbour the Great Western, out of which a new line from Northolt to High Wycombe and Princes Risborough to Grendon Underwood was established. Fearing that a conventional two track route might lead to congestion, a more lavish system was actually constructed with platform loops at the stations and large goods handling facilities. Whilst this elaborate affair was handsomely able to cater for trains from the early part of the century right up to the 1960's, with the then reduction in services and modern colour light signalling much of the track became redundant. Even so British Rail only lifted some of the rails at the time, namely those passing loops at Wembley Stadium, Sudbury and Harrow Road, Sudbury Hill, Denham, Beaconsfield together with the ones on the singled section north from Princes Risborough to Aynho Junction, leaving just Gerrards Cross and High Wycombe with their original trackwork.

Naturally once modernisation had started, with the principal aim being to create a much simplified double track Wycombe line, the first task was to recover any redundant rails. So Gerrards Cross received the treatment first, which allowed British Rail to slew the Down Platform Line across so that it lay parallel to it's Up counterpart. Such action necessitated the building of a new Down platform which is connected to the old one by a short walkway on the trackbed. Several advantages came from this large alteration like the vacating of space for much needed extra car-parking. On the south side of the station, a bay was planned (later replaced by a short siding) so that trains could arrive from London prior to turning back. An identical facility has also been provided to permit an event day only shuttle service to run out of Marylebone to the recently renamed and modernised station serving Wembley's Entertainment Complex and Stadium. Leaving Wembley, Network South East resolved to tackle the problem of Sudbury Hill and Sudbury and Harrow Road using an old Great Central Railway principle, by establishing modern island platforms between the old Up and Down slow lines instead of round them. Fortunately Northolt Park did not possess any through lines and being of simple construction, it needed hardly any work to be done except for a cosmetic face lift.

South Ruislip however was a different kettle of fish, since here some trackwork has been installed in the form of a new trailing connection in the down main line. Rolewise, in order that a peak hour commuter can enjoy faster journey times to the heart of the capital, this station now boasts being the principal interchange with the Central Line, instead of the very much downgraded West Ruislip. To this end West Ruislip has encountered a vast amount of track removal, while the down platform has been extended over the former trackbed of the line through the latter.

From Ruislip onwards, the next few intermediate stations have not required many improvements save for resurfacing of the platforms, revamping of the buildings and the lifting of Beaconsfield's old crossover. After Beaconsfield comes the next largest place namely High Wycombe, which was once an important large station, serving all varieties of traffic. It had been fortunate enough to retain its through lines and some of the old sidings. Naturally with a reduced amount of traffic and state of the art signalling, most of this track has been rendered redundant and lifted. Around the platform area, this entailed removing the up platform line along with the former down main line, leaving the former up main line which by a clever piece of slewing is now serving the platform.

Heading Northwards from High Wycombe along the Chilterns the old Birmingham Main Line skirts through West Wycombe where in recent years a bus rail interchange was planned though these plans came to nought. Unlike West Wycombe, Saunderton has proved to be more fortunate in surviving numerous rail cuts and like its smaller counterparts has benefited from being spruced up.

Unlike Saunderton, Princes Risborough by contrast is very large and has always been important in railway terms having once been a junction for the Watlington and Oxford lines now cut back to Chinnor and Thame respectively. Up until the 20th December 1989 (the same day that Christian Salvesen opened a new Freight Depot at Neasden served by Speedlink Trains from Dover), Chinnor was playing host to a regular flow of coal and cement trains, though faced with the cost of having to replace the life expired wagons used on these trains and the possibility of having to build new unloading and loading facilities, Rugby Cement decided that it would be cheaper to use road haulage instead. After the closure British Rail agreed to let the branch remain in situ a short while longer to allow the infant Chinnor and Princes Risborough Association to attract a sufficient following and of course to raise the necessary finance to purchase the line and acquire a small amount of rolling stock in readiness for running the first trains.

Meanwhile however the future of the Thame branch which had been receiving a regular quota of tank trains destined for the BP Oil Depot also didn't look so rosy at the time with talk of the end of these services not long after the Chinnor operations ceased. In reality a temporary reprieve was issued, then due to the Gulf War hostilities the branch saw much use again, prompting British Rail to include it in the Resignalling Programme, but sadly this was to be too optimistic as it's eventual closure was carried out in March 1991.

Turning to the Banbury line a secure future was assured for this section with the proposed upgrading of the track as far as Aynho Junction and refurbished stations with better parking facilities. Undoubtedly this has not only benefited the Chiltern service but will be of considerable benefit when the small fleet of Class 153 and 156 trains appear in May 1993 to work to Birmingham Snow Hill and Stratford Upon Avon. With the increase in trains if loadings encourage an expansion of the service, it is possible that double track may once again be a feature north of Princes Risborough or at the very least a long passing loop might be provided somewhere near the site of the former Ashendon Junction.

The final line that diverges from Princes Risborough and by far the poor relation on the present day network, also fared really well under the Chiltern Line Scheme, namely the one to Aylesbury which has not only seen the introduction of a much more intensive service but also the revamping of Little Kimble and the complete rebuilding of Monks Risborough station from scratch. It is hoped that the extra services will help generate sufficient revenue to pay for the latter outlay eventually.

At the end of the branch lies Aylesbury where in October 1989, work began with a new facing connection being installed in the single line giving access to the solitary up side siding.

Mind you this event was of little significance to those occurring on the 18th/25 November and 2nd December, when Network South East and Bucks County Council joined forces to sponsor some Shoppers Specials, running from Marylebone to Milton Keynes Central, via High Wycombe and across to Aylesbury, before heading north over the freight only section picking up several passengers from the normally closed Quainton Road and Winslow stations. Each train departed from Marylebone at 8.15am arriving in Milton Keynes at 10.48am and started its journey home at 16.20pm returning just over two and a half hours later to Marylebone at 19.07pm. Some nine hundred passengers enjoyed the option over the three days, prompting the launching of a feasibility study in the following January to see if reopening the Aylesbury to Bletchley line was really feasible. The latter was an unexpected bonus in addition to the slowly progressing works at Aylesbury, where some brand new connections were laid into the Hartwell Siding, as well as into the coal depot for future use.

During February and March 1990, no real major progress occurred, though by the beginning of April, the working party delegated to investigate the Aylesbury to Milton Keynes reopening found that much support existed for the scheme and that with a £1/2 million refurbishment of some first generation DMU's and the provision of platforms at Quainton Road and Winslow, a two hourly service could quite happily be accommodated. However such a plan would depend on Regional Railways filling the alternate hour with one of their envisaged Swindon to Peterborough Expresses, but sadly with the recession there seems scant chance of either plan being carried out before 1995. Thus Aylesbury will have to wait to be a through station officially, but during the Summer of 1990 the disappointment soon faded away as a massive programme of works commenced to remodel the track in the station area which would incorporate some seven sidings on the south side and allow the sharply curved platform four to be swallowed up by an extension of the car park. In addition although most of the character of the station has been retained a new structure has risen up from the platform amidst the Victorian buildings.

Another large building had its construction inaugurated during a small informal ceremony on 24th May 1990, when the first sod was turned for the new Networker Turbo Depot. Fortunately this land formerly a pony field north west of Aylesbury station, had been able to be acquired by British Rail from the Local Council who were keen to accept the abandoned High Street Station area in return for much needed road improvements. This day also saw some three plaques unveiled heralding completion of a £1 million refurbishment of three stations. The latter was divided up into two stages, the first one entailing raising and repaving of the platform surfaces, along with the establishment of new lighting at Wendover and Great Missenden, while the second not long after allowed similar treatment to be carried out at Stoke Mandeville. Simultaneously a full scale major refurbishment commenced on the buildings of the latter along with Wendover and Great Missenden, whose vintage Metropolitan Railway signalbox was now declared to be surplus to requirements, with the station being put under the control of Marylebone.

As far as Marylebone is concerned, it is here probably that the most significant changes have occurred, the first of these being the loss of the successful steam specials. Besides the specials, the regular services saw much recasting to allow two (a fast and a semi fast) trains per hour off peak. During the peak, several trains were scheduled to run non stop from Marylebone to Chorley Wood and Vice Versa. Naturally these improvements in service did little to please the rival Metropolitan who wasted no time at all in throwing down the gauntlet, by announcing that every Amersham train would run straight through to the city instead of stopping at Baker Street. However whilst this was to be of great benefit to travellers, the Metropolitan would never be able to surpass anything that the new Marylebone would offer. Here, after the temporary offices housing British Rail employees had been removed, work commenced with all due speed to excavate the old station service road, ready for the laying of two tracks to serving a new Platform 2 and 3. Meanwhile the contractors were busy creating new toilet facilities for both Men, Women and Disabled people. One major advantage of the latter was that it allowed the old basement to be filled in, after which the concourse could be laid with Network South East's attractive terrazzo flooring. During this process new retail units began to be developed for Travellers Fare and W.H.Smith, but British Rail did not wish these to be the only outlets as previously, so they put several redundant areas of the ground floor up for potential sale or rent for small businesses. Apart from a small area reserved by British Rail for a Customer Service Bureau capable of dealing with train and other enquiries, sufficient accommodation existed on the first floor to create a suite of offices with access via a staircase up from the Old Booking Hall. Purchase of tickets is now done through the Underground Kiosk, it being thought that the retention of the original Great Central facility was unnecessary because the station serves fewer places than it did years ago.

Again within the confines of the main building at Marylebone, a further cause for excitement appeared with the preparation of an Integrated Electronic Control Centre, which would ultimately be the heart of all signalling and communication operations as far as Aylesbury and Banbury. To achieve this aim a special computer was installed with a timetable in its memory that is capable of plotting the routes and allowing the signalman to observe them by means of three television screens, two of which show a large section of line while the third allows any part to be seen very close up. Slave screens were also located at other stations to allow station managers access to any relevant train information. In order to complete this complex system, many new colour light signals were installed along with detector cameras, mirrors and television screens at each station.

By August 1990, considerable progress had been made such that the area of Marylebone's signalling control was extended out to High Wycombe South Box which along with both West Ruislip and Gerrards Cross Boxes, was abolished. At this point a whole mass of antique semaphore signals disappeared after a life of eighty or so years. Whilst the latter showed significant progress, during mid August Network South East resolved to hasten their resignalling plan by temporarily diverting all its trains into Paddington instead of Marylebone. Consequently work could go ahead on completing works at Wembley Hill, Neasden Jncn and Harrow on the Hill. With this Marylebone Box was reduced to a shunt frame while the quaint old Neasden Junction South Box passed into history at the same time as the Civil Engineers reduced the track from double to a single line between Neasden Junction to Neasden South Junction. Hardly any time at all had passed when the last semaphore signals rose and fell for the last time at a London terminus. Marylebone had entered the era of ultra modern signalling. For the ensuing developments in the modernisation of the Chiltern Line, we have to turn to Aylesbury where an ACE Tamper Siding was installed which coincided with the laying of an engineer's siding off the down main line at High Wycombe. Meanwhile a considerable amount of progress was being made on the new £4 million Networker Turbo Maintenance Depot at Aylesbury which was designed to house three tracks two being equipped with service pits while the third was for major jobs. Incorporated in the building the contractors erected a block for a staff of thirty, stores, workshops and some fuelling facilities. Externally the building would be a structure that was streamlined with dimensions of 115m long by 41m wide and 7.5m High, constructed with a stressed metal skin, coated in blue grey and silver, above a red brick foundation wall.

In October 1990 in readiness for the start of the Chiltern Line renaissance, two significant changes occurred, the first being the replacement of Area Manager Peter Leppard who was succeeded by a former Paddington man, Chris Hogan, while the second saw a total recasting of the winter timetable on all routes. However all was not to go smoothly unlike the introduction of the half hourly service the previous May, since difficulty in recruiting sufficient train crews, reliability of the old Class 115 units and weather conditions were all to contribute temporarily to a reduction in the number of services. Although the situation was to continue for a short while, one beneficial thing did happen in the shape of another programme of the Milton Keynes Shoppers Specials which ran on the 3rd and 10th of November with an extra trip on the 17th. Normally these trains were scheduled to run at the start of December, but in 1990 Aylesbury station was closed for resignalling, track and platform work.

Some four days after the last Milton Keynes Shopper Special, the partially refurbished Marylebone Station received a quiet launch by the then secretary of state Cecil Parkinson, who was impressed with the completed concourse. On the platform side, plenty still remained to be done on the platform area with the need to remove number 5 and 6, allowing space to construct a modern office block above an area which was earmarked for stock holding sidings.

Stock Holding sidings were also one of the new features that were seen at Aylesbury where from late November to early December 1990 seven roads were laid in the shape of three trailing sidings off the down side and four facing sidings off the up side main line. Further northwards Platform 4 of the station was being filled in to make extra car parking space, while Platforms 1, 2 and 3 were all upgraded and kitted out with signals for reversible working. Once this work finished Aylesbury South Box closed and the area came under the control of the Marylebone (ME) Signalling Centre from where a track circuit block system operates from Aylesbury to Amersham, an electric token working covers the freight only Aylesbury and Claydon LNE Junction route, leaving the Risborough Branch which runs on a No Signalman Token basis. Commissioning of the new signalling system was also to bring about the end of locomotive 08011 HAVERSHAM the oldest locomotive which performed one final duty namely delivering a BG No 80501 to the infant tourist railway at Chinnor, before being despatched off to Reading.

As the New Year commenced the Modernisation Programme switched to Princes Risborough which was given some further treatment in the form of unnecessary track recoveries and the abolishing of the up main in the station area, which had been used by freight trains and the NSE expresses. These sadly, after years of being cut back from Wolverhampton, then to Birmingham New Street, then to Leamington and finally to Banbury, were not going to be part of the new service pattern, thus much to the disgust of regular passengers on a very cold evening in January the 17.38pm headed by class 47 364 instead of the proposed class 50, departed for Banbury for the very last time.

During February it had been British Rail's intention to complete all the remaining signalling work but on account of the appalling weather, work was deferred until the early part of March when the sections of line between Banbury and High Wycombe and Aylesbury to Princes Risborough were temporarily closed to bring a further stage into operation as far as Bicester North leaving a TCB from here to Aynho Junction controlled by High Wycombe. Consequently High Wycombe, Princes Risborough and Bicester North Signalboxes were abolished, along with Bicester Ground Frame, but its sidings remained. Meanwhile the stock holding sidings were introduced at Marylebone and its fourth platform was reinstated.

The next big event occurred when the first two recently completed Class 165 Networker units No's 001 and 002 emerged from the works at York. Soon these headed away and began numerous trial running schedules between Derby and Leicester from the 10th April. After this they eventually drifted along to the semi complete Aylesbury Depot in readiness for an official ceremony at Marylebone on the 14th May, attended by Malcolm Rifkind who launched the Class 165 Networker Turbo Fleet which heralded the final part of the grand £75 million modernisation plan. Then in a rather short speech Sir Bob Reid very proudly declared "We are today launching our vision of a civilised form of commuting" a statement with a heavy touch of irony considering that Marylebone was to have been closed. Not long after the speeches the assembled dignitaries travelled down via High Wycombe and Princes Risborough (where the old branch to Chinnor was about to be severed) to Aylesbury, so that Mr. Rifkind could open the new maintenance depot.

Sadly within a short time, it was realised that the maintenance depot was not going to see intensive service for some months due to the announcement by BREL that their production forecasts for the Networkers at York had been radically downgraded, such that the full quota of Class 165's would not be delivered until 1992. It was hoped that in the immediate future Chiltern Line would be blessed with 24 two car Networker Turbos, only half its expected complement, to enable inauguration of the new timetable from the beginning of October. As things turned out No 165 006 became the first of the turbos to run in service, when it was used to form the 16.10pm Marylebone to Aylesbury on the 9th September. No 006 was used because 001, 002 and 005 had returned to York for minor attention, while 003, 004 and 007 were on driver training. However just two days later these six other units were finally draughted into service at which point further cars began to appear with a steady flow, the last arriving in early 1992, at around the same time as the last stretch of signalling between Bicester North and Banbury South was switched in, which saw the abolition of Aynho Junction Box and costly "on train" Conductors.

Total cost of modernisation on completion in Spring 1992

New Aylesbury Depot	£ 5 MILLION
New signalling	£12 MILLION
New trains	£46 MILLION
New stations	£10 MILLION
ATP	£ 2 MILLION
TOTAL COST	£75 MILLION

Ironically either part or all of the modernisation that has taken place on The Chiltern Line at vast expense in the last few years will be rendered out of date within a decade, once travellers in the south east see the completion of Crossrail. Crossrail is an exciting project, promoted by the present Government, dedicated to easing road congestion and reducing overcrowding on trains.

Initially it was purely an academic exercise, conceived in 1988 by the Central London Rail Study team, though their ideas caught the attention of Paul Channon, the then Transport Secretary, who commissioned a more in depth examination of the transport system in London and how it related to the commuters actual employment. Quite surprisingly the findings revealed a substantial growth in rush hour travellers of around 19% since 1984 within the central area, which was where a major problem occurred as apart from the Thameslink Kings Cross-Holborn through line, trains descended on termini to unload vast hordes of passengers, who then piled onto an underground system which is in desperate need of expansion to cope with its new found popularity. Undoubtedly one of the reasons for an increase in rail usage has stemmed from a gradual erosion in the number of road users who face restricted vehicle accesses, wheel clamps and slower traffic speeds. Additionally much further traffic has been generated on the main system by commuters heading for the new Light Rail System leading to the Docklands development area. So in light of this along with even further patronage projected, a major rethink of the London transport system was well overdue.

Very soon after the preliminary investigations, an outline paper was produced in January 1989, probably drawing on the success of the Thameslink Kings Cross-Holborn link, advocating the building of further through lines to allow trains to work from one end of the city to the other, thus removing the bottlenecks experienced at British Rail termini and on the underground. It proposed two tunnels under the capital which would house a line running West to East from Paddington to Liverpool Street and one from North to South from Euston to Victoria respectively. Within a couple of months the latter part of Crossrail was abandoned in favour of a new tube line between Chelsea and Hackney, thus connecting the District Line to Wimbledon in the south to the Central Line to Hainault in the north. This decision seems strange since the potential of a Euston to Victoria link would be unlimited, while the envisaged Chelsea to Hackney underground line can only offer a small benefit. However the principal drawback is its cost at some £1.8 billion compared to the £1.5 billion of the West-East plan which has been given priority for construction due to its cheaper price.

Once a definite decision had been taken to build one of the new Crossrail lines, a bill was submitted to Parliament in November 1991, that it is hoped will gain Royal Assent by 1993 ready for works to be started in 1994. These works will include the boring of two 6m diameter single bore tunnels carrying some seven miles of double track with 25kV overhead electrification. This stretch of jointly funded track will contain some five stations, located at Paddington, Bond Street, Tottenham Court Road, Faringdon and Liverpool Street, with possibly a sixth near Holborn all of which would not only be capable of accommodating eight cars of the new 23m rolling stock but also have the potential for being extended should demand require it.

Apart from the works in the capital three suburban lines will be requiring major modification namely Thames, Chiltern and Anglia's London to Shenfield route. By the time construction for Crossrail is ready to begin on the Thames Line, the railway from Paddington as far as Hayes and Harlington West, where a new spur to Heathrow Airport diverges, will have been re-signalled and energised with 25kV electrification. Thus the only major engineering tasks will be to continue electrification and resignalling of the line from Hayes and Harlington West (known as Airport Junction) to Reading. However only the main line, provisionally earmarked for widening with a fifth bi-directionally operated track between Langley and West Drayton, will receive this treatment, thus leaving the High Wycombe, Newbury and Oxford routes and the branches to Greenford, Bourne End and Henley to be diesel operated.

Diesel operation will additionally continue on the Chiltern Line from Banbury, Princes Risborough, High Wycombe and Aylesbury (via Little Kimble) into Marylebone Station, which will be handing its Amersham and Aylesbury services over to Crossrail. Plugging these into Crossrail will require several important alterations to the present infrastructure, namely the building of two new chords to reroute traffic to the Paddington line which will be achieved by running one spur eastwards from Neasden up onto the Cricklewood Freight route, while the second will leave the latter line close to its crossing of the West Coast Main Line, then head westwards skirting round the boundaries of Old Oak Common depot up to where it will join Crossrail at North Pole Junction. Secondly the whole route will have to be electrified as far as Aylesbury on the 25kV system, requiring the raising of bridges, cutting back of foliage and upgrading the track along the route. As soon as construction works commence, all the underground services between Harrow and Amersham will cease operation to allow lifting of the fourth rail power supply along with the recovery of other equipment and the rebuilding of a Rickmansworth station to the east of its present site. Other stations on the Amersham– Aylesbury route will have to have their platforms lengthened. During and after these works the Watford and Uxbridge branches will continue to run as before, but what of the Chesham branch which would seemingly be isolated by Crossrail. Initially no plans had been made for including the branch in the scheme because of the vast expense required to make improvements to the station at Chesham. However at the start of Autumn 1991, after a considerable lobbying by the Federation of Metropolitan Line Users' Committee whose president is Chesham Councillor Gordon Dixon, the British Rail and London Underground bosses finally recognised that Chesham has a significant role to play in Crossrail, so it too will enjoy the proud boast of being able to offer customers a broader choice of travel opportunities such as days out in Essex.

The Essex lines will require little alteration except perhaps a lengthening of the platforms along the routes affected, because a 100% increase in patronage is predicted once Crossrail opens. Presently, the only other work on this section is a resignalling programme being undertaken in connection with the modernisation of Liverpool Street and Crossrail which will enable more trains to be accommodated in the passenger schedule. In all twenty four extra trains will run in the peak time from Essex then once they reach Paddington twelve departures will run out to Reading while the others will head off for Aylesbury. During the off peak hours no definite service pattern has been decided on, although it is more than likely there will be either four or two trains an hour running on an alternating basis like their peak counterparts. In any case with the new Crossrail, Line Potential will exist for any expansion that needs to take place.

To implement the proposed service, using experience gained from the "Networker" project, the Network South East Design Team have managed to come up with a train eminently suitable for both short and medium distance journeys which will be able to provide fast, safe, comfortable travel. Known as the Class 341, the fleet will eventually reach a total of 560, 23 metre long, integral tubular framed, fire proofed aluminum bodied carriages, with rigid anti-deflection or derailment couplings and two strong power units per four car train for increased acceleration. Boarding will be done via three sets of wide doors on each side for fast unloading and loading, which will give the amazed passenger access to an even more impressive interior, incorporating a Real Time Audio Visual Information System, comfortable 2+2 seating separated by an aisle wide enough for wheel chairs, air conditioning, interconnecting gangways, ribbon glazing allowing a clear window for every seat, a black box recorder for incidents and Passenger–Driver /Driver–Signalling Centre communication systems. All being well if the building of these trains and the construction works follow their schedule rigidly, Crossrail may be in action before the year 2000 so beginning another exciting chapter for the Chiltern Line.

Just like days gone by as Metropolitan No 1 stands in the old Brill Branch platform at Quainton Road before backing down onto her train. (Author)

A superb overall view of the preserved station at Quainton Road which is part of the Buckinghamshire Railway Centre. (Author)

In the 1970's and 1980's British Rail used to extend its normal limit of operations on Bank Holidays and Special Event Days in order to convey passengers to the Quainton Railway Centre on "The Quaintonian" which is shown here prior to returning to Aylesbury. (Author)

Probably one of the best examples of carriage restoration in the South of England is this London Chatham and Dover Railway first class carriage which forms part of a four coach vintage train that conveys passengers up and down the West Yard. (Author)

A taste of history on the roast beef special

Mallard — back on duty after a long spell in the National Railway Museum at York.

THE great Mallard — which still holds the world speed record for steam locomotives — is back in business InterCity style.

The thought of such sensation should not upset your Sunday lunch. On the contrary, it could considerably improve it.

The famous Mallard will be coming to London in October to haul the Sunday Roast Beef Special from Marylebone Station to Stratford-on-Avon.

The Mallard is yet another famous name to add to the star-studded array of power, following the Flying Scotsman and Sir Nigel Gresley which have been on duty alongside Clan Line and Sir Lamiel.

The Mallard, built nearly 50 years ago, has been restored to original condition and is on duty after a long spell in the National Railway Museum at York.

The No 4468 Mallard, an LNER Class A4 Pacific, was built in 1938 by Sir Nigel Gresley and on July 7 of that year broke the world speed record for steam traction — 126mph between Grantham and Peterborough on the then LNER line from Kings Cross to Edinburgh.

Attractions

So here is another opportunity to carve yourself a place in history on a day's outing.

Boarding your train at Marylebone just after 10.00, you are served morning coffee on departure and you may enjoy a drink, if you wish, before your three-course lunch.

Upon arrival at Stratford you are free for two hours to roam this fine old town,

Afternoon tea served on the return journey brings to a close an exciting and enjoyable day of sightseeing and nostalgia.

You can even practice your own 'royal wave' to the sightseers who gather at every vantage point to see you — and Mallard — go past in such style.

So why not enjoy InterCity 126 style with Mallard!

OUTWARD		RETURN
d 10.55	MARYLEBONE	a 19.55
	28 miles	
a 11.43 d 11.44p	HIGH WYCOMBE	d 19.05 a 19.03s
	8 miles	
a 12.06c d 12.21	PRINCES RISBOROUGH	18.42 (Pass)
12.39 (Pass)	BICESTER	18.27 (Pass)
	33 miles	
1251 (Pass)	AYNHO JUNCTION	18.15 (Pass)
a 13.00n d 13.20	BANBURY	d 18.06 a 17.36w
13.43 (Pass)	LEAMINGTON SPA	17.04 (Pass)
	22 miles	
a 13.47 d 13.48s	WARWICK	d 16.58 a 16.57p
	13 miles	
a 14.10	STRATFORD UPON AVON	d 16.25

The timing sheet for the 'Shakespeare Limited'.

InterCity

Santa Steam Special

AVAILABLE ON YOUR CHRISTMAS BUFFET TARIFF TODAY

Roast Turkey Roast Chicken Sandwiches
Sausage Rolls Hot Mince Pies

Selection of Soft Drinks Beers
Spirits & Minerals
Tea/Coffee

(BUFFET CAR IS LOCATED BETWEEN CARS C AND D)

Best wishes for Christmas and the New Year

CHRISTMAS STEAM SPECIAL **ADULT**

available only on Special Excursion train
LONDON (MARYLEBONE)

MR HEALY
11.00 dep.

To: **HIGH WYCOMBE** and return

Date of travel: 21 DEC 1986

Coach B Seat 1 ~~Facing~~ Back

NOT VALID UNLESS STAMPED BY ISSUING OFFICE

SPECIAL EVENTS
TRAVEL CENTRE
EUSTON STATION
13 NOV 1986
L. M. REGION
EUSTON
LONDON NW1 2HS

FARE PAID
£8.00
INCLUSIVE OF RESERVED SEAT

№ 1204

Issued subject to the regulations and conditions in the publications and notices of the British Railways Board. Not refundable.

BR4536/1 1444 (M)

The bill of fare for the Santa Steam Special on the 21st December 1986 and a ticket for the same train which ran packed to High Wycombe before returning later in the day. (Author)

A4 LOCOMOTIVE SOCIETY
LIMITED
(A Member of SLOA)

PRESENTS

The Thames-Avon Express

STEAM HAULED TRAIN
OUT OF LONDON

on

Saturday 16th February 1985

featuring:

4498 'Sir Nigel Gresley'

London (Marylebone)—High Wycombe—Banbury—Stratford-upon-Avon—Saltley—Birmingham N.S.—Watford—London (Euston)

SUNDAY 29 DECEMBER 1985

MARYLEBONE TO STRATFORD-UPON-AVON

End the year with a treat for the family; take a nostalgic trip behind a famous steam locomotive, and enjoy a 3-course Roast Beef lunch on the way!

The special train STEAM-HAULED between Marylebone and Banbury in both directions will leave London at 10.30. Morning coffee and lunch will be served en route and Stratford will be reached around 1330. After a leisurely stroll round the delights of the town, you rejoin the train which leaves at 15.30. Optional Afternoon Tea will be served on the return journey and you will be back in London by 18.30.

It is hoped that **'Flying Scotsman'** will haul the All 1st Class train to and from Banbury.

Pullman-type service will be provided to each seat – incorporating Morning Coffee & Luncheon – for an inclusive price of £29.50 (No reduced fares available).

InterCity

THE SHAKESPEARE LTD.

Enjoy a traditional Sunday luncheon on a First Class train with the added romance of steaming through Shakespeare country behind a magnificent historic steam locomotive.

InterCity

Under the vast canopy at Marylebone the Class 115 Diesel Multiple Unit seems like a toy in comparison as it waits to form the 15.10pm departure on the 21 March 1986. Both the canopy above the train and the platforms on the right are no longer while the awning over Platform No 4 has been moved and attached to the columns on the left. (Author)

Looking rather a tight squeeze as the sleek No 4498 Sir Nigel Gresley gently coasts into Banbury station's No 1 Platform with a Thames Avon Express on the 26th January 1985. (Author)

Ancient and Modern combine as No 4498 Sir Nigel Gresley is seen drawing into Stratford Upon Avon with the Thames Avon Express on the 2nd February 1985, while over to the left in Platform No 2 a Diesel Multiple Units waits to depart with an all stations service to Birmingham Moor Street. (Author)

In the early morning of 4th August 1985 the beautifully resplendent Merchant Navy Pacific No 35028 is seen here waiting as she and her train are prepared for the journey ahead to Stratford Upon Avon. (Author)

Captured by the morning sun on the 9th June 1984 a four car Class 115 Diesel Multiple Unit waits in Marylebone's No 4 Platform with the Milton Keynes Shoppers Special. (Author)

A rare shot indeed is this view of the inside of the cab of a Class 115 Diesel Multiple Unit showing the controls and the instrument panel. (Author)

A Class 115 Diesel Multiple Unit waits at Little Kimble whilst working the Fakenham and Dereham Railway Society Bucks and Oxon Freighter Special, to allow enthusiasts to look around the tiny station. Note the destination blind which the driver changed purely for photographic purposes. (Author)

Claydon London and North Eastern Junction Signal Cabin was installed during the Second World War to control a double tracked chord connecting the Oxford to Bletchley line with the Great Central. The chord in revised form is still used but the signal cabin closed a few years after this picture was taken and it has been moved to the infant Swindon and Cricklade Railway Preservation Society. (Author)

CHILTERN LINE
CHRISTMAS SHOPPING TRIPS
to
MILTON KEYNES
by the
"CHILTERN SHOPPER"
special trains
SATURDAYS
16th November
14th December
Adult return fares
Denham, Gerrards Cross, Seer Green, Beaconsfield, High Wycombe, Princes Risborough,
£3.00
Monks Risborough, Little Kimble, Great Missenden, Wendover, Stoke Mandeville Aylesbury, Qainton Road,
£2.50
Winslow,
£1.50
Children HALF fare
NOTE: An additional train from WINSLOW will run on SATURDAY 7th DECEMBER.
Tickets available from your station NOW

Rail campaign gets up steam

A WELL attended meeting at the Old School, Great Missenden, gave support to the campaign against the proposed closure of Marylebone Station and cutbacks in services to Aylesbury and Banbury.

The previous evening, at a private meeting, a pressure group was formed and given the title The Aylesbury and District Rail Passengers Association incorporating the areas covered by the Aylesbury to Amersham and Princes Risborough to Aylesbury lines.

It calls for the retention of the present services, and their expansion.

The chairman of the Missenden meeting, Mr. Leslie Freitag stated that it was necessary for the local people who attended to pass on the message to others, to fight for the retention of Marylebone since two into one would not go.

This latter comment was in respect of a plan to try and load the services of Marylebone into Baker Street.

With regard to the LT plan for electrification north of Amersham he stated: "It would probably be a cheap job and single line with passing places making it less attractive to the consumers".

Indeed he contended it would be absolute folly to use Baker Street as the one terminus since it did not have the capacity. On the other hand Mr. John Healy, who is the A&DRPA press officer, put forward various suggestions for saving Marylebone by saying that there was scope for a Marylebone-Milton Keynes or Marylebone-Birmingham service since these would pass through major industrial centres and shopping centres. Displaced stock from other lines could be draughted on for this purpose. After all the bus services for these areas were far from fast and provided a poor substitute for rail.

The increase in the number of buses meant that the council's developed more roads and hence more cars used them, taking more revenue from the railways. Not only that, roads were more expensive to build and maintain and were detrimental to health. For instance it costs £1 million a mile for a motorway whereas 20 miles of railway can be electrified for this amount.

Mr. Healy summed up by saying that it was necessary to have an integrated road/rail system like a Passenger Transport Executive.

The chairman Mr. Freitag put forward the motion "That the A&DRPA wishes to declare its total opposition to any closure of Marylebone Station and instead seeks improvements rather than cuts".

The audience supported the motion unanimously. It was disclosed during the meeting that 15,000 passengers used the service into Marylebone and they were also eager to stop the proposed plan to sell the station and offices for development as a bus station.

Rail users who wish to join the campaign to retain and upgrade the service can obtain details from The Aylesbury & District Rail Passenger Association by contacting David P. Gray, Alicante, Risborough Road, Terrick, Nr. Aylesbury.

Marylebone passengers form pressure group

Following a public meeting at Aylesbury on Tuesday, Marylebone commuters are setting up a users pressure group and encouraging similar groups at Great Missenden and Wendover.

The feeling of the Civic Centre meeting was that if Marylebone station was closed and sold off, Baker Street station would be unable to cope with the increase in trains.

Commuters either wanted Marylebone to be kept open or special provision to be made so that services did not suffer in any way.

Letters are to be sent to local MPs and the local authorities. The meeting was called by the Railway Development Society.

County to fight station closure

BUCKS County Council is to spearhead a campaign against the threatened closure of Marylebone Station.

Backed by local district councils, it plans to organise a formal protest to the government's transport watchdog body, the Transport Users' Consultative Committee.

In their submissions the councils will say that the British Rail proposals ignore the wishes and convenience of passengers.

Mr. Kenneth Ross, chairman of the County Council's Public Transport subcommittee, said the councils had three aims:

1 Continuation of through services between Aylesbury and London;

2 No significant increase in journey times;

3 No significant decrease in quality or comfort of train services.

British Rail's proposals failed on all three counts.

Aylesbury — Amersham — London services, which would be diverted to Baker Street, would involve a change of trains at Amersham.

The direct service from Aylesbury to Paddington, via High Wycombe, would take 15-30 minutes longer than existing Aylesbury — Marylebone trains.

Plans to put more trains into an already busy Baker Street station could adversely affect peak hour reliability, said Mr. Ross.

He said: "It is clear that British Rail's main reason for suggesting closure to Marylebone is financial. They have given scant attention to the wishes and convenience of passengers."

Call for rethink on Marylebone

THE Aylesbury Society is calling for a "serious re-appraisal" of the proposals to close Marylebone Station.

It has just made its views known in a letter to the London Regional Passengers' Committee, with copies being sent to Aylesbury Vale District Council, M.P. Mr. Timothy Raison, and the Aylesbury and District Railway Passengers' Association.

The letter states that the proposed alternative service between Aylesbury and London appears to be far less convenient for the travelling public, offering a longer journey time.

It would also create a need to change trains, and would provide far less on-train facilities than at present, such as toilets, first-class seating and bicycle space.

EXPANDING

"Aylesbury is a rapidly expanding town, with many national companies having moved their head offices into the town.

"It could therefore be reasonably assumed that a frequent, reliable train service, serving the whole community, is essential to sustain and further this growth which was imposed on the town under the Structure Plan for the South East," states the letter.

The society adds that it wishes to make its observations on behalf of its many "railway-user" members.

New group joins Marylebone fight

AS THE campaign to stave off the threat of closure for Marylebone Station hots up, a new commuter pressure group has been formed in Buckinghamshire to fight for the station's retention.

The new group is the Aylesbury and District Rail Passengers Association and has a recently elected committee which is drawing up battle plans in the war to force British Rail to rethink axing the station.

The transport authority denies, despite widespread opinion to the contrary, that it has any firm intention to shut the London terminal.

The Aylesbury and District Rail Passengers Association came into being at a public meeting at Great Missenden.

It helped to toughen up the opposition against the closure and alleged cut-backs in the services to Aylesbury and Banbury.

The new group is part of the Railway Development Society which is concerned with safeguarding the quality of railway services in general.

The group is looking at the Princes Risborough-Aylesbury and Amersham-Aylesbury lines with the emphasis on improving present services by expansion and development.

John Healy, press officer for the Aylesbury and District Rail Passengers Association, said the chairman of the Great Missenden meeting, Leslie Freitug, thought it necessary for the local people to fight for the retention of Marylebone, since "two into one won't go."

This comment was in respect of a plan to try and load the services of Marylebone into Baker Street.

The meeting declared its total opposition to any closure of Marylebone Station and instead sought improvements rather than cuts.

Shoppers' special in bid to save Marylebone

THE CAMPAIGN to save Marylebone Station from closure is being stepped up by the Aylesbury and District Railway Passengers Association — a pressure group which has supporters in Chiltern district.

As part of its publicity campaign drive aimed at protecting Marylebone — an important terminus for commuters from Buckinghamshire travelling into London — the group has arranged for a special train service to run from the station to Milton Keynes in March.

The Marylebone-Milton Keynes Shopper, as the return service is being called, will stop at selected stations on the Wycombe line to Aylesbury (a connection will be provided with trains from Amersham), where passengers will transfer to another train on a line, which has been closed to regular traffic since 1963, to Milton Keynes.

There passengers can board buses for the shopping centre or a trip round the city.

The scheme is described by a spokesman for the pressure group as an "historical event" which may be the forerunner of many more such services.

The Marylebone-Milton Keynes Shopper will run on Saturday, March 10, with first class, second class and child return fares priced respectively £8, £7, £5.

Further details can be obtained by writing with an sae to the Special Train Co-ordinator, 6 Parton Close, Wendover, where tickets can be supplied.

The association has been analysing what its members want through questionnaires and this information will be sent to the British Railways Board.

The spokesman for the group said that British Rail had now altered its five-year corporate plan to one of three years, which meant that the emphasis now lay more on the association to try and prevent any quick closure of Marylebone.

"Of course one argument for retaining the terminus was proved by recent events at Paddington where the station was put more or less out of action by a train mishap. The Wolverhampton train departed from Marylebone for four days running.

"On all four occasions the concourse at Marylebone was full and the train ran packed, which just goes to prove that if some long-distance trains ran out of Marylebone and Milton Keynes it would encourage greater patronage."

Following initial meetings of the association at Aylesbury and Great Missenden and a recruitment drive, membership is said to have expanded considerably.

BR on line for double trouble

Rail boss to face angry commuters

TWO new initiatives have come in the fight to keep Marylebone Station open — among them a chance to put your views to a top BR boss.

The Marylebone Travellers' Association and the Advertiser have set up a public meeting for October 24. And the MTA is relying on angry commuters to turn up to demonstrate their feelings.

Platform speakers are Chesham and Amersham MP Sir Ian Gilmour and British Rail's deputy director (London and South East Services) Gordon Pettitt.

Beaconsfield MP Tim Smith is hoping Parliamentary commitments will leave him time to attend the meeting.

Both Sir Ian and Mr Smith have come out against the closure plan that was first announced in March this year.

Dr Michael Hargreaves, of the MTA, will chair the meeting which is set for Beaconsfield Town Hall at 8 pm. MTA committee members hope commuters will turn up in force so that Mr Pettitt will return to Rail House having heard at first hand the weight of objections.

And in a second move the MTA is calling for a meeting of commuter groups involved in the campaign to fight the closure.

MTA secretary Tony Shields said: "We believe it is important for associations and local authorities to share information and be aware of each other's actions. This will avoid conflict when arguing against the BR proposals in public."

Letters have gone to other organisations fighting the closure suggesting a meeting and viewing of the various objections sent to the Government body looking into the opposition to closure plans, the London Regional Passenger Committee.

This week a spokesman for the LRPC said objections to the closure were now almost up to the 600 mark and still coming in.

LRPC secretary Nigel Cartwright said although the official closing date for objections was October 8 anything that came in after the date would be looked at.

He didn't forsee any public hearings on the closure until early next year.

Further objections can be sent to the London Regional Passengers' Committee at Room 124, Golden Cross House, 8 Duncannon Street, Strand, London, WC2N 4JF.

Plan to convert rail track for coaches

SHOCK NEWS for Chiltern District rail commuters arrived this week with the announcement that the National Bus Company has drawn up plans to take over Marylebone Station after its predicted closure.

According to the National Bus Company, its bold £10m plan would cut up to an hour off express coach journeys into and out of the capital by using a converted Marylebone track.

That the company should have made such a feasibility study is worrying Buckinghamshire rail pressure groups which now feel he closure is more likely.

The bus company is proposing to convert the first 10 miles of line out of Marylebone into an express coachway, believing that British Rail has long envisaged shutting the terminus.

National, Britain's largest coach operator, intends to end the suburban crawl by inter-city coaches as they leave the motorways to make their way through London traffic.

National says that Marylebone was built as the London terminus of the Great Central Railway main line from the Midlands and North, but in recent years it has handled only suburban services from the Aylesbury and High Wycombe area.

Professor Peter Hall, a leading planning expert from Reading University who has undertaken a feasibility study for the bus company, has explained that the proposed Great Central expressway would take 250,000 coach trips a year off London's choked roads.

Coaches would enter the new coachway at Northolt or Neasden. The converted rail track would then give a clear run through to Marylebone, which would be adapted for coaches and passengers.

And National has emphasised that the coachway would be open to all operators providing scheduled services into London.

National's director of Marketing, Brian Barrett, said: "The cost will be relatively low since the station and track are already there. If BR undertook the transformation in partnership with institutional investors, as it does in its property developments, it would be a money-maker."

Off-peak option

THE Marylebone Travellers' Association is looking to a reduction in off-peak service for the Aylesbury line in order to save the doomed station.

"What we urge all users to do is to consider how to help run their service practically and cost effectively," said association press officer John Healy.

Pointing out there was duplicate British Rail/London Transport service as far as Amersham, he felt this was only necessary in rush hour times.

At off-peak times it would be beneficial for both British Rail and London Transport to run only a Sunday service in order to keep operating costs to a minimum.

In turn, British Rail could then sell off most of the Marylebone site for building leaving just two platforms operational for the Wycombe and Aylesbury trains.

"When it comes to the final crunch in these days of economy and cost effectiveness, we must try to opt for the best practical service from the railways. If the local commuters and users don't make a strong decision and show how much their service means, BR will decide for them, so 'Use it or Lose it'."

The Marylebone Travellers' Association newsletter spoke of the revival of the association with the new threat to the station and the campaign to boost membership through posters and handouts.

Mr Healy spoke of the dangers caused by wild speculation of closures and the guessing game commuters had been playing with BR.

Further information about the Marylebone Travellers' Association can be gained from John Healy, Harehill, Potter Row, Great Missenden.

Support the fight to save Marylebone

THE British Railways Board announced on March 15 that it wants to close Marylebone Station and the approach lines from Harrow and Northolt.

The Marylebone Travellers' Association believes that this proposal would cause severe inconvenience to thousands of British Rail passengers and it has vowed to fight to keep Marylebone open.

The Buckinghamshire Advertiser Series (which covers Denham, Gerrards Cross, The Chalfonts, Beaconsfield, Amersham and Chesham) believes it has a duty to its readers to campaign for the station not to be shut.

The Advertiser has promised full support to the Marylebone Travellers' Association and launched its campaign on March 14 with extensive Page 1 coverage and a leader, part of which declared: "This 85-year-old terminus may well be a tempting nest egg for the hard-pressed money men of the British Railways Board. But the short-term financial advantage of shutting Marylebone is greatly outweighed by the long-term disadvantage of shunting travellers off to the West London terminus of Paddington."

The Advertiser has paid for these leaflets to be printed in an attempt to persuade EVERY Marylebone user to join the association.

BR's customers must speak with one voice if this proposal is to be defeated. The association needs YOUR support if it is to win. Please fill in the coupon TODAY!

SAVE OUR STATION

Advertiser campaign

SAVE OUR STATION

I AM strongly opposed to the plan to close Marylebone Station and divert train services elsewnere. I would like to join the Marylebone Travellers' Association in order to support the campaign to save this station.
I enclose a cheque for £1 to cover one year's subscription to the association, made payable to Marylebone Travellers' Association.

My name is ....

My address is ....

Telephone ....

I travel to Marylebone from

....

Station

Please post this coupon with your cheque to: Save Our Station, Buckinghamshire Advertiser Series, The Vale, Chalfont St Peter, Bucks SL9 9ST.

CLOSURE OF MARYLEBONE

CLOSURE OF **MARYLEBONE STATION** WILL AFFECT YOU!

HOW WOULD YOU TRAVEL THEN? HELP PREVENT A CLOSURE!

THE MARYLEBONE TRAVELLERS ASSOCIATION IS FIGHTING AND NEEDS YOUR SUPPORT

REMEMBER THE 1970 FIGHT – THE THREAT WAS THE SAME, WE FOUGHT AND WE WON!

JOIN THE ASSOCIATION – SEE OVER FOR DETAILS

HOW WILL IT AFFECT YOU?

May 7, 1986

Safe at last

By Jacki Davis

JUBILANT commuters are celebrating victory following the decision by British Rail to scrap its plan to close Marylebone Station.

The news has been greeted with delight by rail users who joined the Advertiser's two-year battle to save the station from the axe.

Last Wednesday's historic announcement by British Rail came as a surprise to campaigners, who had been expecting a date for the long-awaited public inquiry into the planned closure to be announced, rather than news of a reprieve.

Rail chiefs say they have decided to keep Marylebone open because the underground could not cope with the extra traffic which would have been diverted onto the Metropolitan line if Marylebone had closed.

A spokesman for London Underground said the number of passengers using the Metropolitan Line into Baker Street at peak times had increased by 15 per cent since 1983.

He said the increased traffic was largely due to the success of the Travelcard and Capitalcard in persuading more commuters to travel to work by underground.

He said: "The passengers that would have switched from British Rail, added to the substantial growth in traffic we have had, would have made for very crowded and uncomfortable trains."

The closure proposals, announced in March 1984, involved diverting trains from High Wycombe, through Beaconsfield and Gerrards Cross, to Paddington and transferring passengers from the Aylesbury line to the London Regional Transport services at Amersham.

Marylebone would have been sold and the stations at Wembley Complex, Sudbury and Harrow Road, Sudbury Hill and Northolt Park would have closed.

Delighted members of the Marylebone Travellers' Association, which mounted a widespread publicity campaign against closure, have hailed British Rail's decision as a victory for common sense.

Secretary Tony Shields, of Lower Road, Chalfont St Peter, said: "Obviously we are very pleased. It has been two years of sweat, but we are very pleased that British Rail has withdrawn its proposals."

Mr Shields believes the main reason for BR's decision was its recognition that the Metropolitan Line could not cope with the extra traffic from Marylebone.

But he added: "I am sure that they recognised the strength of feeling against closure. We have had two public meetings ourselves, which were both packed, and there has been a tremendous demonstration of public support for the campaign against British Rail."

And association chairman Dr Michael Hargreaves, of Grove Road, Beaconsfield, said: "It never was a very sensible proposal and one always welcomes it when one's opponents see the light."

The jubilation among members of the Marylebone Travellers' Association was shared by Chesham and Amersham MP Sir Ian Gilmour, who was a keen supporter of the campaign to save the station.

● Tony Shields — very pleased.

He said: "Obviously I am extremely pleased. It seems to me that they have realised that their figures were wrong and ours were right."

Sir Ian said he was surprised by British Rail's announcement: "I had a nasty feeling that British Rail had made up its mind before it all started, although, at the public meeting I addressed it did strike me that they had a very bad case."

British Rail is now considering reducing the number of platforms at Marylebone and selling part of the site to developers. But its spokesman said the board was committed to improving facilities at the station and encouraging more commuters to use it.

He said: "We are really going to go ahead now and promote Marylebone in a big way. We are thinking about new trains in the future, improving the station and generally brightening up the service."

This commitment was welcomed by Mr Shields, who praised British Rail for the steps it had already taken to smarten up the station and attract more customers by introducing special steam train journeys on Sundays.

Mr Shields said the Marylebone Travellers' Association was due to meet British Rail on Wednesday, May 14, to discuss steps to revitalise Marylebone and improve services from the station.

And Dr Hargreaves said he hoped British Rail would take steps to improve its services. He said: "They have been running it down, but there was not much we could do about it because they said they were going to close it anyway. Now they are not going to close it for the foreseeable future, they had better get their fingers out."

And he added: "If we get a half-hourly service, then nobody can complain."

Timetable

THE battle to save Marylebone Station from the axe began more than two years ago.

1984

● March: British Rail formally announces intention to close Marylebone.

● July: British Rail accused of underhand tactics by publishing the official closure notices during the summer holiday season.

● August: Six-week deadline for comments on the closure is extended by a month. Protests flood in.

● September: British Rail finally publishes timetable for after the closure.

● October: Chesham and Amersham MP Sir Ian Gilmour accuses British Rail chiefs of defensive secrecy at a packed public meeting.

● December: Save Our Station campaigners pack into public meeting. MTA launches a fighting fund to pay barrister at public inquiry.

1985

● February: News that BR has postponed public hearing due to take place at the end of the month is greeted with cheers. Delay follows admission by BR that its orignal figures were wrong.

● May: Public inquiry is postponed for second time because of legal wrangle over cross-examination of witnesses. Brent Borough Council appeals against High Court ruling upholding inquiry procedures.

● November: Court of Appeal dismisses Brent's appeal, but Brent warns it will seek permission to appeal to House of Lords.

1986

● April: Legal wrangle ends with Brent being refused leave to appeal to House of Lords. Report produced to London Regional Transport says Marylebone would not be suitable site for coach station.

Better deal for rail travellers

THE reprieve for Marylebone Station is to have far-reaching effects on the quality and efficiency of rail travel in this area starting from next Monday.

In addition to immediate timetable changes — which involve the introduction of an earlier direct train from Aylesbury to Marylebone in the mornings and the re-timing of some early evening services — British Rail has long-term plans for spending out on improvements to track, rolling stock, signalling and stations.

The changes to the Marylebone to High Wycombe and Banbury service and the Marylebone to Aylesbury service are designed to make the service more reliable, say British Rail.

As one of the Network South East managers, Mr. John Oxley, explained:

"Now that the long-term future of Marylebone and its train services are assured, it is important that we provide the Chiltern lines with a more economic and efficient service.

"The changes will match train capacity more closely with demand, speed up journeys for the growing number of longer distance commuters, and reflect the trend towards earlier commuting," he said.

An earlier direct train will leave Aylesbury for Marylebone at 0637, and Marylebone for Aylesbury at 0640.

Three later commuter trains from Aylesbury — the 0820, 0840 and 0910 — have been combined into two leaving the town at 0825 and 0900.

And two early evening commuter trains at 1610 and 1640 have been retimed to 1600 and 1628, otherwise the Marylebone to Aylesbury peak hour service remains unchanged.

Theatre goers will welcome a slight change to weekend services with the 2338 Marylebone to Aylesbury via High Wycombe Saturday train departing at 2355 as it does during the week.

Commuters from Banbury, Bicester, Princes Risborough and High Wycombe will benefit from a new train running non-stop between High Wycombe and Marylebone arriving at 0922 and departing Marylebone in the evening at 1725.

In the long term, an ambitious programme of improvements to track and signalling is planned which will speed up journey times and prepare for the replacement of the present rolling stock and introduction of modern diesel units, planned for 1990.

Details of the improvement programme are still under wraps but a British Rail spokesman says Aylesbury has had a taste of what is to come as it was one of the first stations along the line to be repainted, re-signed and generally spruced up.

Marylebone at the other end is also getting a face-lift following its reprieve, and the emphasis is switching to clearing the lines of rubbish.

Foliage and trees in areas where they are likely to cause a danger to train drivers and users are being cut back and verges and embankments are being tidied.

£MILLION RAIL INVESTMENT FOR BUCKINGHAMSHIRE

ONE MILLION POUNDS IS TO BE SPENT ON MODERNISING THREE BUCKINGHAMSHIRE STATIONS THIS SUMMER. Network SouthEast has already started work at Wendover and Great Missenden stations raising and repaving the platform surfaces. Pictured here is John Powell, Chairman, Buckinghamshire County Council, Public Transport Committee tapping the last paviour into place at Great Missenden station. Over £300K has already been spent on this initial stage of the project which includes renewing the lighting on the platforms at both stations.

Contracts have just been let for the next stage of this major project. Work will begin this month on the station buildings at Stoke Mandeville, Wendover and Great Missenden which are to be refurbished at a cost of over £600K. And as part of this second stage the platforms and lighting at Stoke Mandeville will be brought up to the standard of Wendover and Great Missenden.

Peter Leppard, Area Manager, Chiltern, said: 'Over the next two years the heart and the face of the Chiltern Line will be renewed. Trains, stations, track and signalling are all being modernised step-by-step. Work on these three stations should be completed early in 1990 and at the beginning of 1991 I hope to see the first of the new Network Turbo trains coming into service. The future is looking very bright for Chiltern Line at the moment.'

Milton Keynes bears all the hallmarks of the best kind of modern stations since it is not only blessed with two fast and two slow lines each served by platforms with adequate facilities and a bay which hopefully will receive trains from Marylebone one day, but also it boasts a concourse and booking hall of generous proportions at road level. (Author)

Outside Marylebone showing the attractive archway giving access to the concourse which has been lovingly cleaned and restored. Note the letters carved onto the crest in the centre and those on the gates which bear witness to the station's original owner. (Ann Priest)

After years of neglect the concourse at Marylebone has now been transformed with a new marble floor, cleaning of the buildings, a larger walk in W.H.Smiths and some new shops, a joint Underground and Network South East and a totally modern ticket gate with train information screens. (Ann Priest)

The time is 12.24pm and a few passengers head through the ticket gate to board the 12.37pm Marylebone to Aylesbury working which is sitting in Platform No 4. (Ann Priest)

Just inside the ticket gate at Marylebone station looking northwards from Platform No 3 and 4. On the left a train is waiting to form an Aylesbury working while that on the right will be heading for High Wycombe. (Ann Priest)

The spacious interior of one of the ultra modern Class 165 Networker Turbos showing the brightly lit compartment with its banks of individual seats, spacious luggage racks and a walk through facility making it a more welcome variation on its predecessor. (Ann Priest)

An overall view of the canopy at Marylebone in its present form after the demolition of the third span. Once this was complete the roadway under the second span was dug out to provide two new platforms. (Ann Priest)

Under the canopy at Marylebone with a splendid line up of Networker Turbos in evidence on the 20th May 1992. Note the floral displays and the extra lighting both of which are intended to brighten up the platforms. (Ann Priest)

A multitude of railway activity at Neasden with the busy London Transport depot on the right while in the centre are the main lines. Over on the left where a twin siding railhead has been established for Christian Salvesen, a huge depot once stood belonging to British Rail. (Ann Priest)

Wembley Stadium station which is the alighting point for the Stadium, Arena and Conference centre where Class 165 No 010 and No 014 are waiting to depart for High Wycombe and Marylebone respectively on 10th June 1992. (Ann Priest)

An unidentified Class 47 rumbles through Wembley Stadium station with an engineer's train bound for the sidings at Princes Risborough. (Ann Priest)

As a passenger boards an unidentified Class 165 on 25th July 1992, the signal is set for it to take the Northolt route to Marylebone which curves away sharply on the left. Straight ahead the line leads to Paddington. (Ann Priest)

Having just left the flying junction at Northolt, a Class 165 Networker Turbo No 007 gently eases into South Ruislip with the 12.40pm Marylebone to Banbury working. (Ann Priest)

A rather attractive close up picture of Class 165 No 011 waiting patiently at South Ruislip for some custom while on a run to High Wycombe. (Ann Priest)

West Ruislip station at the end of the down line platform for the Wycombe line trains showing one of the station's old signalboxes, while in the background a Central Line train waits to depart for the City and Essex. (Ann Priest)

West Ruislip is the interchange between British Rail and The Underground on the High Wycombe route and here it is possible to see a Central Line station on the right while on the right while that on the left with the enlarged down platform serves Marylebone and Banbury trains. (Ann Priest)

Saunderton station with a new Network South East Class 165 Turbo two coach train drawing into Platform No 2 with the 12.20pm ex Marylebone to Princes Risborough service. (Ann Priest)

Sadly due to the removal of a large amount of trackwork and the down platform at Princes Risborough whose trains now call at the up side, the station has much unutilised space and looks somewhat of an eyesore. It is a real pity that British Rail did not rebuild the down platform and demolish the up platform thus creating further space for car parking or a potential development site. (Ann Priest)

In this photograph there is a fine pair of modern junction signals in evidence and a platform mirror though the scene is dominated by the Networker Turbo Class No 165 011 which is just drawing into Princes Risborough from Banbury with a Marylebone service. On the left is the fast line and the two tracks which led to the Chinnor and Thame Branches. (Ann Priest)

Without any doubt the North Signalbox which outlived its southern counterpart by just over two decades is the most handsome feature of Princes Risborough and indeed it once had a massive 100 lever signalling frame the length of a cricket pitch controlling every point and signal around the platforms and those on the first stages of the lines to Watlington, Aylesbury, Oxford and Banbury. In recent years with track rationalisation, only half the box was needed, then in 1991 it closed altogether. (Ann Priest)

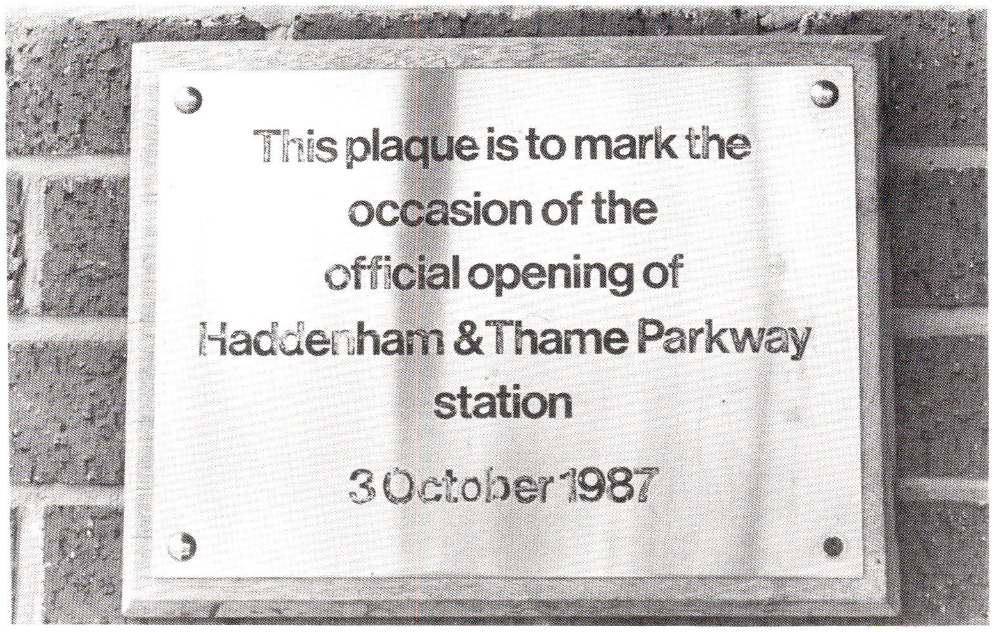

The plaque commemorating the opening of a new station at Haddenham some twenty four years after the original one closed. To help recoup some of the cost a wider catchment area is being tapped hence the addition of Thame. (Ann Priest)

Haddenham and Thame Parkway station platform sits on the site of the former down main line to Birmingham. In this view a service is seen just about to leave for Marylebone. (Ann Priest)

A rather basic but nevertheless very functional reception building at Haddenham and Thame Parkway station which is situated above platform level. (Ann Priest)

A truly impressive view of the driver at work whilst on a run from Banbury to Marylebone with one of the modern Class 165 units. Note the well laid out, user friendly control panel. (Ann Priest)

At Banbury station a rather smart looking Guard waves his flag to give the right away to a Manchester bound express. (Ann Priest)

Once upon the principal train of the Somerset and Dorset Railway was the Pines Express which in recent years has been resurrected as a High Speed Train working and here it is seen heading through Banbury on the Southbound run from Manchester Picadilly to Bournemouth Central. (Ann Priest)

Little Kimble from the platform side showing the really spartan facilities afforded to passengers in the 1990's. Note the foreshortened canopy which hardly protects the traveller from the rain. (Ann Priest)

The approach to Little Kimble station showing the small car park and the gateway which now gives access to the platform. Many years ago any potential passengers would enter through the building on the right which was sold off by British Rail and converted into an attractive private dwelling. (Ann Priest)

Chorley Wood Station on the 1st June 1991 at the south end with an eight car Metropolitan A60 set gently heading away to Baker Street with the 14.49pm ex Amersham. Note the old Metropolitan style signalbox. (Ann Priest)

Tranquility and peacefulness prevails at Chorley Wood but not for long as the indicator board hanging from the old ornate canopy proudly proclaims the arrival of a train for Baker Street and the city. (Ann Priest)

At 16.08pm on Wednesday the 17th June 1992 one of the new Networker Turbo trains is seen sitting in the up platform at Chalfont and Latimer with the 15.39pm ex Aylesbury to Marylebone working (Ann Priest)

Looking from the Marylebone end of the Down Platform our expert photographer has managed to record an interesting picture of Chalfont and Latimer looking towards Amersham, showing the attractive nature of the original part of the up platform, the rather austere extensions to the canopies and the Chesham Branch bay in the background on the right. (Ann Priest)

From 1971 no steam trains had actually travelled on London Underground but the 1989 centenary of the construction of the Chesham Branch changed that as Metropolitan No 1 and Great Western No 9466 seen here posing with the Mayor of Chesham and a London Transport Inspector, shuttled from Watford to Chesham and back with a series of specials on the weekends of the 1st/2nd and 8th/9th July. (Ann Priest)

A rather fascinating photograph of Amersham station at the southern end showing the Signalbox at the back, Platform No's 1 and 2 which are both kitted out for bi-directional working and serve Aylesbury trains and all arrivals from Baker Street, leaving Platform No 3 nearest to the camera which plays host to both Metropolitan and Network South East workings bound for London. (Ann Priest)

An overall view of Amersham station with a Networker Turbo sitting in platform No 1 whilst waiting to depart with the 10.19am service to Aylesbury on the 7th May 1992. (Ann Priest)

Amersham station as viewed from the north end of Platform No 3, showing the Metropolitan stabling sidings which were added during electrification in the late 1950's. Note the train in the up side siding which is waiting to form the 14.48pm Aldgate working. (Ann Priest)

The south end of Great Missenden station showing both the signalbox which is sadly defunct but remains as a reminder to the past and the trailing crossover which is now rarely used. (Author)

One of the old order in the shape of the only named Class 115 Unit Aylesbury College waiting to depart from Wendover in late Summer 1991 with a Marylebone train. (Author)

Wendover station on the 17th July 1991 at 10.47am looking towards Aylesbury with a good handful of passengers having assembled for the 10.49am departure to Marylebone. (Author)

An overall view of Stoke Mandeville showing the splendid restoration that has been done to spruce up the original Metropolitan Buildings, Canopy and Footbridge. Note the attractive platform tiling. (Author)

If only all stations were like this on Network South East complete with their own landscaped garden. This picture shows Stoke Mandeville with its beautifully tended lawn and bushes. Note the statue in the foreground which was erected in recognition of the way a former stationmaster loving cared for his charge. (Author)

The down platform at Stoke Mandeville where the facilities are really basic as can be seen from the bus shelter type waiting area. However compared to previous years the would be passenger is now more well informed with a handy clock and timetable. (Author)

This view depicts an incredible transformation as at one time the goods yard dominated the scene on the right, but now it has been developed by Safeways. On the left on what was formerly a recreation ground, Network South East have established their Maintenance Depot to serve the Networker Turbos. (Ann Priest)

Faster, larger trains will take the strain

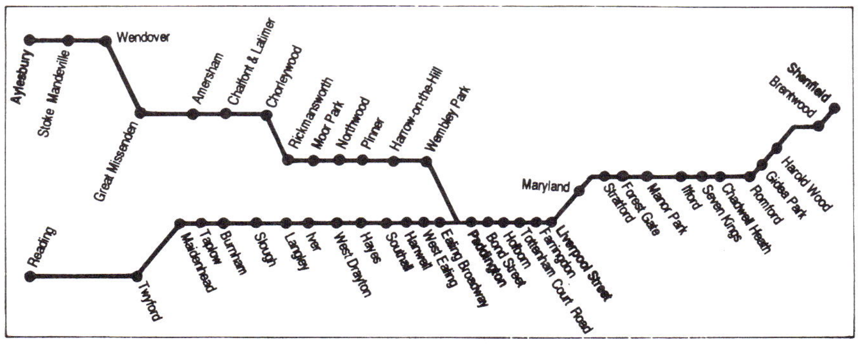

Speed lines: The proposed cross London rail link promises major benefits for Aylesbury travellers, says ALEC BROWN

FASTER journey times and greater comfort have been promised for Aylesbury travellers when the new cross-London electrified rail link is up and running at the end of the decade.

A Bill to build the cross London link has just been introduced in the House of Commons but details of what it would contain were first given to Bucks Herald readers in May 1989 when the plans were leaked.

The CrossRail service, which is a joint project between Network South East and the London Underground, will mean journeys from Aylesbury, Stoke Mandeville and Wendover will come in to Paddington instead of Marylebone.

In order to do this CrossRail will take over the existing Chiltern line from Aylesbury to Amersham and the existing Metropolitan line from Amersham to Wembley Park. The route will then continue along a little used freight line south to the BR Old Oak Common depot from whereabout a mile of new track will have to be built to link with the main line to Paddington.

The route will continue through central London on a new underground line and then go east out of London as far as Shenfield in Essex.There will also be links from Paddington to Reading and on.

Trains from Aylesbury via Risborough, and the Banbury to London line will continue to use Marylebone.

A mock-up of one of the CrossRail coaches, which will be the largest ever to run underground in Britain, is on display at Paddington.

Transport minister Roger Freeman unveiled the coach last week and people are being asked for their views.

A spokesman for CrossRail said journey times will be cut dramatically.

The train from Aylesbury now takes 64 minutes to reach Marylebone; CrossRail will reach Paddington in 53 minutes.

But there are even greater savings for journeys into central London. Shoppers going to the proposed CrossRail station at Tottenham Court Road will be able to get there in 58 instead of 77 minutes, while businessmen going to Liverpool Street will make it in 65 instead of 85 minutes.

There will also be four trains an hour at peak times — twice the current number.

The Aylesbury line was chosen as part of the route because it passed through areas where many commuters lived, said the spokesman.

Many Central Line tube travellers were found to come from areas served by the Aylesbury line, so it was also hoped to ease congestion on that busy tube route.

"Marylebone will continue to have a role for the other services that leave there at the moment. There are no proposals to make any changes except on the Aylesbury-Metropolitan line," he said.

This means Princes Risborough and Haddenham trains though not on the CrossRail route, will benefit from quicker journeys in central London.

The new trains will by powered by overhead electricity cables, so some tracks may have to be lowered or bridges raised, and some platforms extended to take the larger trains.

But the bulk of the building work will be in central London, he said.

The current £74 million Chiltern Line improvements will not be wasted, he said, as improvements carried out to signalling, tracks and stations would have been needed anyway, and the newly introduced Turbo trains will be used elsewhere in Network South East, he said.

CrossRail was born out of the Central London Rail Study report in 1988/89, which proposed solutions to the expected 15 per cent growth in passenger numbers by 2001.

Transport secretary Malcolm Rifkind gave the go-ahead in October and a Bill to get the powers necessary to build the line was submitted to Parliament in November.

If consent is given, building work could start by 1994 and the line could be working by

CROSSRAIL JOURNEY TIMES FROM THE WEST DIRECT TO CENTRAL LONDON

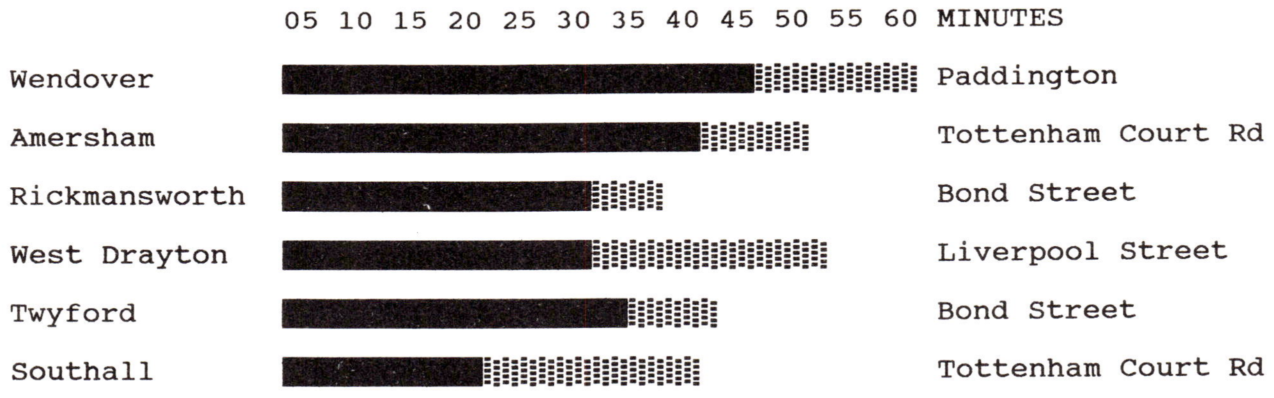

TRAINS PER HOUR EAST BOUND		TRAINS PER HOUR WEST BOUND	
FROM READING	04	FROM SHENFIELD	04
FROM MAIDENHEAD	06	FROM BRENTWOOD	08
FROM SLOUGH	10	FROM GIDEA PARK	18
FROM HAYES AND HARLINGTON	10	FROM CHADWELL HEATH	22
FROM EALING BROADWAY	14	FROM ILFORD	24
FROM AYLESBURY	04		
FROM AMERSHAM	10		